Algorithmic Culture

Algorithmic Culture

How Big Data and Artificial Intelligence Are Transforming Everyday Life

Edited by
Stefka Hristova, Soonkwan Hong, and
Jennifer Daryl Slack

LEXINGTON BOOKS
Lanham • Boulder • New York • London

Published by Lexington Books
An imprint of The Rowman & Littlefield Publishing Group, Inc.
4501 Forbes Boulevard, Suite 200, Lanham, Maryland 20706
www.rowman.com

6 Tinworth Street, London SE11 5AL, United Kingdom

British Library Cataloguing in Publication Information Available

Library of Congress Cataloging-in-Publication Data

Names: Hristova, Stefka, 1977- editor. | Hong, Soonkwan, 1974- editor. | Slack, Jennifer Daryl, editor.
Title: Algorithmic culture : how big data and artificial intelligence are transforming everyday life / edited by Stefka Hristova, Soonkwan Hong, and Jennifer Daryl Slack.
Description: Lanham : Lexington Books, 2020. | Includes bibliographical references and index. | Summary: "This book explores the complex ways in which algorithms and big data are reshaping everyday culture, while at the same time perpetuating inequality and intersectional discrimination. It situates issues of humanity, identity, and culture in relation to free will, surveillance, capitalism, neoliberalism, consumerism, solipsism, and creativity"—Provided by publisher.
Identifiers: LCCN 2020039737 (print) | LCCN 2020039738 (ebook) | ISBN 9781793635730 (cloth) | ISBN 9781793635747 (epub)
Subjects: LCSH: Algorithms—Social aspects. | Information society. | Technological innovations—Social aspects. | Big data—Social aspects.
Classification: LCC HM851 .A5556 2020 (print) | LCC HM851 (ebook) | DDC 303.48/33—dc23
LC record available at https://lccn.loc.gov/2020039737
LC ebook record available at https://lccn.loc.gov/2020039738

Contents

Introduction

In the Presence of Algorithms

Stefka Hristova, Soonkwan Hong, and Jennifer Daryl Slack[1]

Being in the presence of algorithms poses questions about the ways in which computational mechanisms have come to permeate everyday life. Before we start to untangle their influence on identity, collectivity, and culture, we must grapple with the ways they manifest as an opaque technology that insists on having an omniscient view over both our datified selves and our physical bodies. Langdon Winner aptly writes that

> At issue is the claim that the machines, structures, and systems of modern material culture can be accurately judged not only by their contributions to efficiency and productivity, not merely for their positive and negative environmental side effects, but also for the ways in which they can embody specific forms of power and authority.[2]

Following Winner's insight that artifacts do indeed have politics, we are interested in the ways algorithms shape relations of power in everyday culture. Driven by private ventures, algorithms currently operate as proprietary black boxes. As Frank Pasquale has shown, algorithms are both pervasive and elusive.[3] While their presence is felt, their mechanics are absent from the public view. Following the appeal by Judge Damon Keith, "Democracy dies behind closed doors," we argue that exposing their presence and mode of operation has become a task of both cultural and political importance. What is at stake in illuminating the black box of algorithms is the preservation of democracy itself. Points of critical intervention can be identified by highlighting the biases and power dynamics articulated in an algorithmic culture. Algorithmic technology is still developing, and we have an opportunity to situate its practices in a historical context, to interrogate its current manifestations, and to image the future differently.

The work of algorithms demands constant unconsented monitoring of the social. Their presence thus manifests as omniscient surveillance of the ascendant "data self" and the inscription of inequity onto the descendent communal self. These data selves represent algorithmic collective types, types that John Cheney-Lippold has eloquently named "measurable types."[4] The omniscient view of algorithms presents a "big picture" view of the social. As Ruha Benjamin powerfully writes, "If surveillance treats people as 'surface,' then countering this form of violent exposure can entail listening deeply to the everyday encounters of those who are forcibly watched."[5] Vertical engagement with the lived on-the-ground experiences and the enforced digital profiling mechanisms of the algorithm on the cloud becomes an important strategy for understanding the cultural landscape. This vertical visuality is important as often in the constitution of the collective data self-identities on the margins of the social become lost and even erased.

Algorithms have an uneasy relationship with human agency. Their presence also implies the absence of human meddling. Algorithms are seen as faster, better, and stronger than people, and their "abilities" have translated humanness as a "disability." Human beings are prone to mistakes, messiness, and moodiness. Cultures are processual, dynamic, and affective. Culture and algorithms function in opposition to one another. Algorithmic culture becomes a site where their incompatibility leads to impasses as well as productive points of disconnect.

Algorithms are hardly the first and will likely not be the last of technological innovations that aim to harness, tame, calculate, and predict culture in the name of a neoliberal, market-driven world order. They are part of a larger genealogy of technologies of quantification that emerged in the nineteenth century with disciplines such as phrenology and anthropometrics. They are equally invested in the cultural shift toward reducing people to numbers as well as to the raising binary logics of computing and digitization. Therefore, in tracing the more recent development of computing, it is important to explore the ways algorithmic culture stands apart from what emerged in the early 2000s as "digital culture."

FROM DIGITAL TO ALGORITHMIC CULTURE

Algorithmic culture continues and contributes to an important transition in the way information is organized and knowledge is produced. It diverges from the paradigm of digital culture where the digital was primarily centered around the emergence of digital media formats delivered in a distributed online network: the World Wide Web. This network consisted of "pages" or documents. While this process rendered the physical structure of media into computer code, the artifact was preserved as a digital document. Lev Manovich emerged as one of the pioneer theorists of this then "new" media landscape. In his seminal text *The Language of New Media*, he lucidly theorized "new" meaning "digital" media as having five distinct principles.[6]

Digital media objects, according to Manovich, are first of all "digital code" and thus exist as "numerical representation."[7] The implications here are significant precursors for the articulation of a subsequent algorithmic media and algorithmic culture, respectively. Rendered as numerical representations, digital objects can be described mathematically and further subjected to algorithmic manipulation. This process of algorithmic manipulation has become intensified and indeed ubiquitous in an algorithmic culture. In rendering an object digital, two important processes take place: first, a document is sampled at regular intervals creating a grid of pixels to represent a digital image; second, each sample is assigned a numerical value.[8] This technique is inherently cartographic, as continuous analog reality becomes reduced to a set of coordinates positioned on a spatially fixed grid. At this stage, reality is understood as continuous while the digital object is a set of coordinates. This process, as the fifth principle of "new" or "digital" media will show, is subsequently reversed and reflected back to culture as space and territory become themselves distilled and understood as a set of GPS coordinates, or as what Lisa Parks has called "data-lands."[9]

The second principle of new media, according to Manovich, is "modularity."[10] Here, new media are understood to have a "fractal structure" where smaller elements are aggregated into large scale objects while at the same time preserving their independence as complete, coherent, comprehensive objects. This fractal quality of digital media allows for multiple ways of recombining documents; for example one can have still images, sounds, and video embedded in hypertext-based websites. The act of reorganizing media has indeed been seen as one of the core characteristics of digital culture. Mizuko Ito, danah boyd, and Henry Jenkins, among others, have aptly described digital culture as "participatory" culture, as a culture of remix.[11] This principle is compromised in algorithmic culture, as the mechanism of remix is relegated to machine learning and thus obscured from the view of the ordinary user. Algorithmic culture restricts active user participation.

The third principle of new media postulates that the first two principles (numeric coding and modular structure) allow for the automation of "many operations involved in media creation, manipulation and access."[12] The implications of this feature entail the reconfiguration of human agency and the creative process. In the context of digital media, the filters and presets of popular imaging software packages such as *Adobe Photoshop* became iconic representations of this third principle.[13] The click of a button can transform a color image into a stylized black-and-white visual. Algorithmic media are indeed heavily reliant on this process of the automation of creativity. Algorithms have produced music, poetry, as well as video commercials. Beyond creativity, the algorithmic automation is currently extended past the realms of media making and media remixing. Currently, algorithms calculate human emotions, predispositions for purchase, student success, employability, and health by parsing speech patterns and facial expressions, etc. into a set of seemingly equivocal data.[14]

The fourth principle of new media engages with the concept of "variability," where digital media objects are never finite objects but rather exist in "different, potentially infinite versions."[15] New media objects are seen as "mutable" or "liquid." This property of mutability of media has come to foster a culture of update, which Wendy Hui Kyong Chun argues has become habitual.[16] Just as users update their status on social media and websites update their content in order to stay relevant or "alive," updates are often associated with habitual behaviors and are incremental rather than substantive. The process of variability in an algorithmic culture is applied to cultural practice as well as to the articulation of what constitutes humans and humanity. Whereas in a digital culture one updates their *Facebook* status, in an algorithmic environment one's status is automatically updated based on a set of data points. Self-driving cars, for example, continuously track a "subject" in a frame in order to determine the persistence of the subject and to further categorize the subject as a pedestrian.[17]

The fifth, and final principle of new media is especially significant as it articulates the impact that digital media logics have on the reshaping of culture. "Transcoding," according to Manovich, is the act of translating something into another format. The transcoding of media into culture has resulted in the computerization of culture. Here "cultural categories and concepts are substituted, on the level of meaning and/or language, by new ones that derive from computer ontology, epistemology, and pragmatics."[18] In other words, computer ontology becomes the prevalent cultural ontology. The digitization of culture extends beyond media in that it transforms vast areas of culture well beyond media into digital and, further, algorithmic objects. For example, one's iris becomes reduced to over 240 iris data points by digital biometric technology, rendering what was known in photographic and cinematic terms as an extreme close up of an eye as no longer significant as an image document but rather as a distilled data set. Understood as an algorithmic object, the digital image of an iris becomes constituted and evaluated solely as numeric data through a set of computational principles.[19] The totality of the image is indeed irrelevant and one might say obliterated. While the making of the digital image has always and already been subjected to a biased algorithm, in an algorithmic culture the algorithm extends beyond the image in order to surveil and make calculations about human and nonhuman subjects alike.

In the context of digital culture, the digital media paradigm has influenced a participatory culture of remix and frequent social media updates. Algorithmic culture articulates a particular type of technological culture where data and not documents drive decisions and further decisions are made increasingly by machine learning algorithms on the basis of probability and risk. Further, these decisions extend well beyond the scope of media into the social world. This paradigm shift has been theorized as the reduction of culture to a network akin to the Internet of Things.[20] In algorithmic culture, then, it is increasingly important to understand the ways algorithmic structures are shaping everyday life through recognition, calculation, automation, and prediction.

WHAT IS AN ALGORITHM?

In her book, *Hello World: Being Human in the Age of Algorithms*, Hannah Fry presents a highly versatile definition of an algorithm. According to Fry, an algorithm is "the invisible pieces of code that form the gears and cogs of the modern machine."[21] Fry classifies algorithms based on function in four major categories: (1) prioritization or the making of an ordered list (*Google Search, Netflix*); (2) classification or the picking of a category (advertising and measurable types); (3) association or the finding of links (dating algorithms, *Amazon*'s recommendations); (4) filtering or isolating what's important (separate signal from noise).[22] Further, she argues that based on paradigms, algorithms can be divided into two main groups: (1) rule-based algorithms where instructions are constructed by a human and are direct and unambiguous (the logic of the cake recipe) and (2) machine learning algorithms which are inspired by how living creatures learn.[23]

Algorithms thus vary in their logic and purpose. They work together in a fractal manner to create larger automation structures based on the principles of "machine learning." As Meredith Broussard has aptly noted, this learning is limited to the ways the machine "can improve at its programming, routine, automated tasks."[24] This "learning" can be further classified into three categories: supervised, unsupervised, and reinforcement.[25] What is key here is that by anchoring algorithms in machine learning, a paradigm of training, "learning," and predicting is set in motion as "the algorithms are coupled with variables to create a mathematical model."[26] Algorithms train on data sets in order to articulate rules that are then applied to new data sets. In an algorithmic culture, everyday life has itself become a set of data to be regulated by algorithms.

ALGORITHMS AT WORK

Langdon Winner's insight into reconfiguring technological assemblages is particularly relevant to the algorithmic landscape;

> Technologies identified as problematic would be taken apart with the express aim of studying their interconnectedness and their relationship to human need. Prominent structures of apparatus, technique, and organization would be, temporarily at least, disconnected and made unworkable in order to provide an opportunity to learn what they are doing to humankind.[27]

As computing and computational logics are becoming a ubiquitous and pervasive thread of the social fabric, big data and machine learning algorithms are transforming all aspects of everyday life. Algorithms quietly but powerfully contribute to shaping what is possible and what is not, what matters and what does not, what we are becoming, and what we leave behind. So much of this work has been happening below the level of awareness and accountability; rapid change is occurring with insufficient attention to political, ethical, and cultural ramifications. It is important to move beyond simply exploring the articulation of algorithms and big data into a more focused discussion about the ways human bias embedded in both technologies reproduces inequality and oppression. Algorithms shape the ways we perceive traditional categories such as race, class, gender, and sexuality by literally reinventing their symbolic structure in the space of the digital and by directly impacting the lived experience of so many.

The structures of inequality that algorithms create have been exposed and theorized increasingly over the last decade. Ruha Benjamin has detailed the complex ways algorithms produce "coded inequity" where "Blackness is coded as criminal."[28] Safiya Noble has brought light to the ways search engines such as *Google* have reinforced racism. Catherine D'Ignazio and Lauren Klein have delved deep into the ways big data and machine learning reproduce gender inequality as "data-driven decision-making can be just as easily used to amplify the inequities already entrenched in public life."[29] Virginia Eubanks has powerfully demonstrated the ways police algorithms target the poor. Cathy O'Neil has written extensively on the impact big data and algorithmic processes have on our ability to obtain loans, secure insurance, enroll in college, and become employable unevenly by highlighting the hurdles and disadvantages immigrants face.[30] This volume extends the critical work on big data and algorithms by exploring the ways these technologies function as catalysts exacerbating inequality in a global, neoliberal, surveillance-driven culture.

CHAPTER OVERVIEWS

Chapter 1

In this chapter, Slack and Hristova make an argument for advancing analysis in terms of *algorithmic culture* rather than in terms of algorithms, as if the technical achievements of these mathematical processes and their application to complex tasks were significant in and of themselves. They argue that foregrounding algorithmic culture demands addressing the *connections* that constitute what matters most about algorithms: their integration in practices, policies, politics, economics, and everyday life with consequential political, ethical, and affective significance. While it is *possible* to talk about these elements separately, the more pressing challenge is to recognize that and how they are inextricably linked in ways that matter. The story that needs to be told involves (at the very least) math, probability, technology, marketing, consumption, power, policymaking, ethics, privacy, art, and what it means to be human. It is a story that requires integrating the knowledge and practice of data scientists, engineers, social scientists, cultural theorists, artists, and philosophers. It is only in the connections among these that matters of power and consequence can be addressed with the attention they deserve.

Chapter 2

Hong provides a critique of the current state of the new socio-technical environment, relevant practices, and the political economy of the algorithms in relation to consumer culture. He argues that algorithms sequester consumers' lived experiences and promote fetishization of what they are and what they can offer in the market. As a result, the confusion consumers experience between symbols and signs connotes that consumption is highly choreographed. Hong uses the *Matrix* metaphor to describe the engineering of reality and truth based on the well-known practice McDonaldization. For Hong, consumers' "data-mindedness" creates a new normal that lacks idiosyncratic meanings and values as critical ingredients for a multiplicity of identities. He identifies three stances consumers take in response to algorithmic consumer culture: celebratory, strategic, and critical. He suggests that taking up genealogy as an attitude and critique assists in analyzing and theorizing the convoluted power relations in the market where a subject-object dialectic is frequently and openly reversed.

Chapter 3

Algorithmic technology is infused with bias. The first place where this bias can be countered is in the making of the algorithm where a homogeneous task force creates an Echo Chamber in which diversity is severely lacking. Sasha Constanza-Chock has written extensively about the importance of implementing a design justice framework in order to challenge inequity. She issues a call to "dismantle the matrix of domination and challenge intersectional, structural inequality" and argues that this process "requires more than a recognition that employment diversity increases capitalist profitability."[31] This notion of diversifying the process of algorithmic making is further complicated by Ushinish Sengupta's argument that beyond diversity, the notion of monoculture needs to be taken seriously. Sengupta argues that "trying to address algorithm development cultural issues by diversifying the employee pool, but maintaining the same cultural base, will not solve deeply rooted culturally dependent issues." In other words, the training of diverse groups of individuals in algorithmic design and development, without examining the cultural basis of algorithm development will likely perpetuate inequitable results. Diversity in design is situated in relation to diversity of culture. This point runs counter to assumptions that the software development industry is "acultural," independent of culture, or often "post-cultural," transcending cultural differences in developing the future. This chapter argues that software development and particularly algorithm development is deeply rooted in narrow cultural practices. Moreover, current cultural practices of algorithm development exclude genuine input from a variety of stakeholders, while simultaneously the negative impacts of algorithm development, such as algorithmic bias, are experienced primarily by communities and individuals who are not meaningfully involved in the development of the algorithms that affect them.

Chapter 4

This chapter stages algorithmic culture not in relation to surveillance and control in the totalitarian sense but rather as "attaining efficiency in staging a kind of chaos." Chakraborty's analysis rests on Gilles Châtelet, who offers an illuminating perspective that juxtaposes freedom to the histories of individual freedoms, real and theoretical. These could be the freedom to buy and sell in the market or the notion of absolute freedom in the Hobbesian state of nature. Such accounts of freedom are subject to various degrees of chaos. Chakraborty explores two theories that situate individual freedoms jostling with each other: thermocracy and neurocracy. He illuminates the relevance of the ways Châtelet contrasts the benign chaos of the free market with the destructive chaos of the Hobbesian state of nature to characterize the illusion of freedom that defines market democracies. This comparison allows him to comprehend the chaos of opinions that is being staged in the Post-Truth context. Chaos tends to come with the promise of self-organization as well as the anxiety of anarchy. The uneasy stability of the opinion-market is the only possible bargain of an equilibrium. For Chakraborty, "self-organization is the teleological impulse organizing algorithmic culture."

Chapter 5

Algorithms rely on the processing of big data. Big data, articulated by John Cheney-Lippold, "represents a shift towards an algorithmic production of knowledge that is regarded as more true, and more efficient, than nonalgorithmic, non-big-data forms."[32] This knowledge is gained through what Reka Patricia Gal terms "hacking of the social" where big data is harnessed under the "hacker" logic in service of consumerism and capitalism. In outlining the transition away from hacking for liberation and toward hacking for social control, Gal centers on the "turn to the dark side" of the hacker, the programmer, and the engineer into data capitalist or data entrepreneur as a microcosm for the historical transition to neoliberalism. Gal focuses on the figure of the hacker in order to highlight the ways in which the Hacker Ethic of the 1960s has been turned against itself in surveillance capitalism, failing its core commitment that "information should be free." Hacking in the context of surveillance capitalism involves not the tweaking of code but rather the reprogramming of the social fabric based on big data. Gal guides the reader through two historical moments in which the hacker ethic becomes harnessed for surveillance capitalism: first, she explores how hacking principles were instrumentalized in capital's attempts to re-start accumulation after the 1970s crisis on new technological grounds, eventually resulting in the "dot com bubble" and, second, she analyzes the turn to capturing and profiting from social behavior in today's Big Data Age.

Chapter 6

Big data is mined and harnessed in order to articulate "big-data algorithmic identities" based on statistics."[33] These identities get further wrapped in what Wendy Hui Kyong Chun and Sarah Friedland articulate as "imaginary bubbles of privacy."[34] In her chapter, Hristova takes on the audience of Amanda Todd's suicide video in order to examine the implications of the construction of a "mourning" measurable type. She compares the ways social media have created consumer driven filter bubbles around suicide and death in the context of a Web 2.0 and Web 3.0 ecologies respectively. She argues that whereas social media in the Web 2.0 era sought hardwired playlists and explicit dependence on commercial sponsors, in the age of algorithmic media or Web 3.0, affect became the media currency where content and commerce flowed interchangeably. The stakes of this shift are significant as viewers become caught in the inner logic of the algorithm: the algorithm articulates the viewer as a measurable type and then proceeds to deliver content that is both homogeneous and centered onto the perceived interests/classification of the viewer. Algorithms, in general, thus exhibit solipsistic homogeneity that manifests in the form of bubbles of privacy. In thinking about the ways death and mourning operate algorithmically in social media, she argues that algorithms of mourning not only lack but also actively erase and, at the same time, capitalize on the cessation of humanity as they deliver what Jacque Derrida has theorized as autoimmunitary violence.

Chapter 7

Algorithms are increasingly reshaping social identity not only as a category in the present but also as a marker of one's heritage, ancestry, of one's past. As Ruha Benjamin writes, "The quest for scientific tools to determine ancestry and arbitrate group membership continues aspace toward a variety of political and biomedical fields."[35] These "technologies of identity [further] redefine the social categories of identity."[36] In thinking through the ancestry services offered by companies like *23 and Me, Ancestory.com*, and *National Geographic*, Amanda Girard argues that in this process groups with limited representation become erased. Genetic testing has become seemingly open and available to the public through companies like *23 and Me* and *Ancestory.com*, although much of the information is based on algorithms that the companies will not release to the public. Multiple reports of inaccuracy when evaluating twins have made headlines; however, these companies continue to sell the idea of easy genetic testing to provide a historical profile of identity. The evidence suggests the use of a machine learning algorithm may be leaving some groups out of any findings—especially those groups who may already have limited numbers. This erasure is two pronged: it is driven by the machine learning algorithm and by the amount of data acquired. As Girard astutely demonstrates, "The size of a database is important because the information you receive about your ancestry is only as good as the amount of information that the company already has on file." She stresses that "the cultural ramifications of lost or erased populations are much farther reaching" as they redefine identity not only in terms of data but also as a fundamental social category.

Chapter 8

Algorithmic bias also extends the cultural bias into new technological landscapes. Benjamin has detailed photography's historical investment in the production and reproduction of whiteness masked as neutrality. In the transition to digital technologies, "social and political factors continue to fashion computer-generated images [in a way in which] race is not only digitized but heightened and accorded greater value."[37] Joel Beatty dive deeply into the biased algorithms that have structured digital photography and digital cameras more specifically in order to illustrate how the "human" in the loop construct already predisposes a particular biased construction of humanity rather than a neutral, universal, and hence "standard" apparatus. Beatty argues that a critical understanding of where bias occurs along this chain is of high value for an algorithmic culture that includes engineers, policymakers, technical communicators, and a consuming public alike. For this purpose, this chapter investigates the historical process of designing algorithms for the digital color input/output system beginning with the articulation of the CIE Color Space in 1930s and poses critical questions about the "algorithmic logic" that contributes to the emergent cultural and subjective landscape of digitally colored imagery. Beatty locates bias in the nexus of cultural convenience and argues that "the development of digital cameras along with its system of algorithms, databases and technical processes, was driven by the underlying value of convenience" and calls on consumers to be "critical consumers of digital cameras [as] we have a vital role to play" as they "import human biases into autonomous and semi-autonomous loops that our culture is responsible for."

Chapter 9

One of the most pertinent features of algorithms is the displacement of human agency in key social processes. In countering the processes of automation, Broussard has called for the systematic and intentional inclusion of human agency and the subsequent creation of "humans-in-the-loop systems."[38] The automation of culture and art posit yet another important way algorithms have transformed human experience. As James MacDevitt eloquently writes, algorithms emerge as "significant, if not supreme, shapers of human experience" as they "no longer just metaphorically replicate, or even functionally replace human actions; they now indirectly influence, and in some cases even directly supervise, those activities as well, often without the conscious knowledge of the human beings involved." MacDevit analyzes the displacement of the human in the creative process by focusing on the generative adversarial networks (GANs) machine learning technique developed in 2014 by Ian Goodfellow. This chapter traces two parallel and interwoven trajectories that highlight the growing connections between contemporary art and algorithmic processes. First, it explores the evolution of the automated generation of visual imagery by computational machines from pioneering computer art engineers (such as A. Michael Noll, Manfred Mohr, Frieder Nake, Georg Nees, and Vera Molnár) to popularly accessible visual programming languages (such as *Processing*, developed by Casey Raes) to artificial intelligence systems that self-generate completely novel visual imagery (such as Harold Cohen's *AARON*, Pindar Van Arman's *Cloudpainter*, and the recent work of Mario Klingermann and the art collective known as Obvious). The second trajectory focuses on conceptually oriented and process-based practices including those that existed coeval with the dawn of computational imagery (such as Fluxus artists Yoko Ono and George Brecht and Conceptual Artist Sol LeWitt). From Fluxus and Conceptual Art, the chapter moves on to examine systems based artists (such as Roy Ascott, Alison Knowles, Channa Horwitz, Steve Roden, and Dawn Ertl), as well as media artists whose meta-works parody the algorithmic-based interfaces that now govern the distribution of ubiquitous content (such as Jason Salavon and Natalie Bookchin) and hacktivist artists who create conceptual works directly within the algorithmically coded environments used by their viewers (such as Owen Mundy, whose online project *I Know Where Your Cat Lives* purposefully surfaces the locational metadata hidden within uploaded photographs, and Philipp Schmitt's *Declassifier*, which highlights the machine learning algorithms that process and order those very images). Lastly, the trajectory tracing the narrative of algorithmic conceptualism revisits projects that work with artificial intelligences, this time specifically exploring pieces that critique the social and political implications of these semi-autonomous creative tools (including works by Stephanie Dinkins, Mary Flanagan, Trevor Paglen, and Tega Brain).

NOTES

1. We thank the Institute for Policy, Ethics, and Culture at Michigan Technological University for the support to develop the discussion on algorithmic culture.

2. Winner, Langdon. "Do Artifacts Have Politics?" *Daedalus*, Vol. 109. No 1 (Winter 1980), 121.

3. Pasquale, Frank. *Black Box Society: The Secret Algorithms that Control Money and Information*. Cambridge: Harvard University Press (2015).

4. Cheney-Lippold. *We Are Data: Algorithms and the Making of Our Digital Selves*. New York: New York University Press (2017), 57.

5. Ruha, Benjamin. *Race after Technology*. Cambridge, UK/Malden, MA: Polity Press (2019), 128.

6. Manovich, Lev. *The Language of New Media*. Cambridge, MA/London, England: MIT Press (2001).

7. Ibid., 27.

8. "Pixel" https://www.figma.com/dictionary/pixel/ (Accessed May, 2020).

9. Parks, Lisa. "Vertical Mediation and U.S. Drone War in the Horn of Africa." In *Life in the Age of Drone Warfare*. Edited by Lisa Parks and Caren Kaplan. Durham and London: Duke University Press (2017), 134–158.

10. Manovich, *Language of New Media*, 30.

11. Jenkins, Henry, Mizuko Ito, and Danah Boyd. *Participatory Culture in a Networked Era*. Cambridge, UK/Malden, MA: Polity (2016).
12. Ibid., 32.
13. Adobe, "Convert A Color Image to Black and White" (May 21, 2020).
14. Bogen, M., and A. Rieke. "Help Wanted." Upworthy.com (December 2018).
15. Ibid., 36.
16. Chun, Wendy Hui Kyong, *Updating to Remain the Same*. Cambridge, MA: The MIT Press (2016).
17. Dwivedi, Priya. "Tracking Pedestrians for Self Driving Cars." *Towards Data Science* (April 13, 2017).
18. Manovich, *Language of New Media*, 45.
19. Electronic Frontier Foundation, "Street Surveillance: Iris Recognition" (2019)."
20. Burgess, Matt, "What Is the Internet of Things." *Wired.co.uk.* (February 16, 2018).
21. Fry, Hannah. *Hello World: Being Human in the Age of Algorithms*. New York: W.W. Norton & Company (2019), 2.
22. Ibid., 8–9.
23. Ibid., 10–11.
24. Broussard, Meredith, *Artificial Unintelligence: How Computers Misunderstand the World*. Cambridge, MA: The MIT Press (2018), 89.
25. Ibid., 93.
26. Ibid., 94.
27. Winner, Langdon. *Autonomous Technology: Technics-out-of-Control as a Theme in Political Thought*. Cambridge and London: The MIT Press, 1977., 330.
28. Benjamin, *Race after Technology*, 124.
29. D'Ignazio, Catherine, and Lauren Klein. *Data Feminism*. Cambridge, MA: The MIT Press, 2020. https://bookbook.pubpub.org/pub/dgv16l22/release/6
30. O'Neil, "Weapons of Math Destruction."
31. Constanza-Chock, Sasha. *Design Justice: Community-Led Practices to Build the Worlds We need*. Cambridge, MA: The MIT Press, 2020. https://design-justice.pubpub.org/pub/cfohnud7/release/1
32. Cheney-Lippold, *We Are Data*, 57.
33. Ibid., 58.
34. Chun, Wendy Hui Kyong and Friedland Sarah. "Habits of Leaking: Of Sluts and Network Cards." *Differences*, 2015, 26(2), 4.
35. Benjamin, *Race after Technology*132.
36. Ibid.
37. Ibid., 100.
38. Broussard, A*rtificial Unintelligence*, 177.

BIBLIOGRAPHY

Adobe, "Convert a Color Image to Black and White" (2020). https://helpx.adobe.com/photoshop/using/convert-color-image-black-white.html (Accessed May, 2020).

Benjamin, Ruha. *Race after Technology: Abolitionist Tools for the New Jim Code*. Cambridge, UK/Malden, MA: Polity, 2019.

Bogen, Miranda, and Rieke, Aaron. "Help Wanted: An examination of Algorithms, Equity, and Bias." *Upworthy.com*, 2018. https://www.upturn.org/static/reports/2018/hiring-algorithms/files/Upturn%20--%20Help%20Wanted%20-%20An%20Exploration%20of%20Hiring%20Algorithms,%20Equity%20and%20Bias.pdf (Accessed June, 2020).

Broussard, Meredith. *Artificial Unintelligence: How Computers Misunderstand the World*. Cambridge, MA: The MIT Press, 2018.

Burgess, Matt. "What Is the Internet of Things." *Wired.co.uk.*, 2018. https://www.wired.co.uk/article/internet-of-things-what-is-explained-iot (Accessed May, 2020).

Cheney-Lippold, John. *We Are Data: Algorithms and the Making of Our Digital Selves*. New York: New York University Press, 2017.

Chun, Wendy Hui Kyong. *Updating to Remain the Same: Habitual New Media*. Cambridge, MA: The MIT Press, 2016.

Chun, Wendy Hui Kyong and Friedland Sarah. "Habits of Leaking: Of Sluts and Network Cards." *Differences*, 2015, 26(2), 1–28.

Constanza-Chock, Sasha. *Design Justice: Community-Led Practices to Build the Worlds We need*. Cambridge, MA: The MIT Press, 2020.

D'Ignazio, Catherine, and Lauren Klein. *Data Feminism*. Cambridge, MA: The MIT Press, 2020.

Dwivedi, Priya. "Tracking Pedestrians for Self Driving Cars." *Towards Data Science*, 2017. https://towardsdatascience.com/tracking-pedestrians-for-self-driving-cars-ccf588acd170 (Accessed May, 2020).

Electronic Frontier Foundation. "Street Surveillance: Iris Recognition" (2019). https://www.eff.org/pages/iris-recognition (Accessed May, 2020).

Eubanks, Virginia. *Automating Inequality: How High-Tech Tools Profile, Police, and Punish the Poor*. Boston: St. Martin's Press, 2018.

Fry, Hannah. *Hello World: Being Human in the Age of Algorithms*. New York: W.W. Norton & Company, 2019.

Jenkins, Henry, Mizuko Ito, and Danah Boyd. *Participatory Culture in a Networked Era*. Cambridge, UK/Malden, MA: Polity, 2016.

Manovich, Lev. *The Language of New Media*. MIT Press: Cambridge, MA/London, England, 2001.

Noble, Safiya. *Algorithms of Oppression: How Search Engines Reinforce Racism*. New York, NY: New York University Press, 2018.

O'Neil, Cathy. *Weapons of Math Destruction*. New York, NY: Crown, 2016.

Parks, Lisa. "Vertical Mediation and U.S. Drone War in the Horn of Africa." In *Life in the Age of Drone Warfare*, edited by Lisa Parks and Caren Kaplan. Durham and London: Duke University Press, 2017, pp. 134–158.

Pasquale, Frank. *Black Box Society: The Secret Algorithms that Control Money and Information*. Cambridge: Harvard University Press, 2015.

"Pixel" https://www.figma.com/dictionary/pixel/ (Accessed May, 2020).

Ruha, Benjamin. *Race after Technology*. Cambridge, UK/Malden, MA: Polity Press, 2019.

Winner, Langdon. *Autonomous Technology: Technics-out-of-Control as a Theme in Political Thought*. Cambridge and London: The MIT Press, 1977.

———. "Do Artifacts have Politics?" *Daedalus*, 1980, 109(1), 121–136.

Chapter 1

Why We Need the Concept of Algorithmic Culture

Jennifer Daryl Slack and Stefka Hristova

The story of algorithmic culture has become increasingly important to tell. Some of our most venerated colleagues in university are those in computer and data science who study artificial intelligence, big data, and machine learning, all of which are algorithmically driven. Some of the wealthiest and most powerful people on the planet are those who have harnessed the use of algorithms to amplify media messages and influence politics. The newest forms of communication and art are algorithmically driven, and all of us live in a world where algorithms increasingly shape everyday life. These everyday realities include matters as diverse as the behavior of the stock market; how people are hired, evaluated, and fired; how college admission decisions are made; how we study and understand climate change; the possible futures of self-driving vehicles; emerging practices of policing, incarceration, and parole; whether or not you are recognized by voice or visual recognition systems; how drones target deadly kills; how healthcare is organized and delivered; how news is filtered; what and how we have been guided to consume; and the ways we find love and maintain friendships. There is, in fact, very little in our daily lives untouched by algorithms. From the mundane to life and death matters, functioning mostly in the background, algorithms quietly but powerfully and unforgivingly contribute to shaping what is possible and what is not, what matters and what does not, who thrives and who does not, what the world is becoming, and what we leave behind. As U.S. National Security Administration (NSA) systems administrator and whistleblower Edward Snowden put it recently:

> All of these things [devices and everyday realities] are increasingly being created and programmed and decided by algorithms, and those algorithms are fueled by precisely the innocent data that our devices are creating all of the time: constantly, invisibly, quietly, right now.[1]

The collection of data, the algorithmically collected activity records, the "meta data," "tells the whole story," Snowden continues:

> These activity records are being created and shared and collected and intercepted constantly by companies and governments. And ultimately it means as they sell these, as they trade these, as they make their businesses on the backs of these records, what they are selling is not information. What they are selling is us. They are selling our future. They are selling our past. They are selling our history, our identity, and ultimately they are stealing our power and making our stories work for them.[2]

These virtually invisible processes and the myriad links that get us from the technical design of mathematical algorithms to what we are calling "algorithmic culture," demand our attention in order to grasp the significance of what Snowden means by "stealing our power and making our stories work for them" and enable us to resist or acquiesce. How ultimately are our lives changing as algorithmic governance proliferates? Why should we care? How can and should we respond?

We urge thinking in terms of *algorithmic culture* rather than in terms of algorithms, as if the technical achievements of these mathematical processes and their application to complex tasks were significant in and of themselves. Foregrounding algorithmic culture demands addressing the *connections* that constitute what matters most about algorithms: their integration in practices, policies, politics, economics, and everyday life with consequential political, ethical, and affective significance. While it is *possible* to talk about these elements separately, the more pressing challenge is to recognize that and how they are inextricably linked in ways that matter. The story that needs to be told involves (at the very least) math, probability, technology, marketing, consumption, power, policymaking, ethics, privacy, and what it means to be human. It is a story that requires integrating the knowledge and practice of data scientists, engineers, social scientists, cultural theorists, artists, and philosophers. For it is only in the connections among these that matters of power and consequence can be addressed with the attention they deserve.

Our approach has some affinity to the concept of the "structure of feeling" in which Raymond Williams described metaphorically the relationship between elements and the whole (the culture) as a relationship between precipitate and solution: "We learn each element as a precipitate, but in the living experience of the time every element was in solution, an inseparable part of a complex whole."[3] A structure of feeling is constituted of connections among "thinking and feeling which is indeed social and material" and includes myriad "forms of social life."[4] It is a "complex of developments" that includes technologies, social processes, forms of social organization, institutions, *and* feelings and beliefs.[5] The "lived experience" for Williams consists of contradictions, settlements, and changing relationships visible in a momentary and changing "set of emphases and responses within . . . determining limits."[6] So while such a structure is not fixed, uniform, or permanent, it is a set of relationships that exhibits some tenacity that has consequences.

Just as we have learned to study technology not as a mere artifact but as "*articulations among the physical arrangements of matter, typically labeled technologies, and a range of contingently related practices, representations, experiences, and affects*," algorithms are best understood as complex arrangements of math, matter, and related practices, representations, experiences, affects, and effects.[7] Recognizing layered complexity as a necessary theoretical position is only, however, the beginning of the task. It is far too easy to acknowledge complexity and proceed to study a phenomenon *as if* it were simpler. The question we must ask is, "How do we move from recognizing the complexity of context within which algorithms are developed, implemented, and effective to studying that context?" Again, insisting on foregrounding algorithmic culture is a way to work toward that goal. That insistence is the most obvious answer to "why we need algorithm culture" but then demands addressing the question, "What is algorithm culture?" Indeed, the answers to both questions are inextricably linked.

We did not invent the term "algorithmic culture." It emerges first as a theoretical nod, an inchoate sense that algorithms matter, in the title of Alexander Galloway's 2006 book, *Gaming: Essays on Algorithmic Culture*.[8] Ted Striphas's 2015 article "Algorithmic Culture" begins to bring the term into a tighter focus to highlight the way that "human beings have been delegating the work of culture—the sorting, classifying and hierarchizing of people, places, objects and ideas—to data-intensive computational processes."[9] In a footnote, Striphas makes a case for thinking about algorithms consistent with the arguments we make here:

> Algorithms are best conceived as "socio-technical assemblages" joining together the human and the nonhuman, the cultural and the computational. Having said that, a key stake in algorithmic culture is the automation of cultural decision-making processes, taking the latter significantly out of people's hands.[10]

Striphas's research traces the conditions "out of which a data-driven algorithmic culture has developed . . . to offer a preliminary sense of what 'it' is," he provides a crucial historical reading of "how" algorithms as a form of information have taken on a significant role in cultural decision making.[11]

The approach we take here is meant to augment this "preliminary sense," to identify the work of the "precipitates," the elements that constitute the solution and how they connect to give substance to the term "algorithmic culture."

Beyond recognizing that the design and implementation of specific machine algorithms have significant effects, the concept "algorithmic culture" draws attention to the reality that culture is increasingly explained by, responsive to, and shaped in and by the pervasive work of algorithms. Algorithmic culture exhibits a distinctive circularity, an enfolding really, where algorithms are increasingly used to "explain" culture and culture is increasingly crafted "to become" algorithmic. This enfolding can be mapped in four cascading movements that respond to, create, and account for the fabric of algorithmic culture. These movements, variously recognized in the emerging literature on algorithms, entail (1) the work of selection, transformation, and bias in the design of algorithms, (2) the process of machine learning, in which the "learning" escapes human design, (3) the negotiation of power and control among users and algorithms, including the transformation of skilled worker and machine tender into algorithmic troubleshooter, and (4) the reshaping of culture into the producer of computational logics, including the creation of a subjective landscape where computational thinking and machine logic are normalized as commonsense in navigating the cultural terrain. Emergent scholarship on algorithms most often addresses the first two of these movements. The challenge of "algorithmic culture" is to recognize the ways these four movements articulate—resist, augment, and work in relation—thereby demanding that we rethink work, governance, education, economics, consumption, surveillance, privacy, and so on, the relations among them, the resulting constraints imposed, and a range of possibilities unleashed.

ALGORITHMIC DESIGN: SELECTION, TRANSFORMATION, AND BIAS

An algorithm in its most stripped-down, reductionist iteration is little more than a procedure for describing and executing an operation with a predictable outcome: a set of steps or rules for getting something done. A recipe for cooking food is often used as an example to illustrate a simple—noncomputerized—algorithm. An algorithm specifies the ingredients to be selected and the order and process by which they are combined. Recipes are tested and adjusted to ensure they consistently deliver a reliable outcome. This straightforward description carries with it a profoundly misleading sense of innocence because even the simplest recipe entails significant cultural biases and effects. The specific ingredients selected (their cost and availability), their reduction and translation into a language recipe readers understand (language, systems of measurement), the technologies used (the tools required and available for use), the distribution of the recipe (in a recipe online or in a physical cookbook), and the circumstances of implementing a recipe (in a real cooking situation with the time and capacity to cook) make a recipe accessible and useful for some and inaccessible and useless for others. Each of these stages entails the introduction of significant selection and bias, with correspondingly significant effects. Together they result in significantly different outcomes/objects/identities. Someone's grocery store mac 'n cheese is not the same in quality, taste, and health outcomes as a gourmet magazine's mac 'n cheese, even though both are the products of a deceptively similar algorithm. The cultural (if not mathematical) devil, of course, is in the algorithmic details. Even this simplest of examples demonstrates that algorithms entail cultural choices and processes that not only include and exclude but prioritize and value the lives and circumstances of some over others with unequal and discriminatory outcomes.

When scholars talk about algorithms, they are typically referring to mathematical and computer-assisted processes to achieve some desired end. Striphas described algorithms as "a set of mathematical procedures whose purpose is to expose some truth or tendency about the world."[12] When data scientists talk about algorithms, they are typically referring to computer-assisted computational processes arranged to achieve a desired end in which the path to the goals[13]—the processes of selection, reduction to machine language, availability, and implementation that introduce significant cultural effects—are largely irrelevant and typically overlooked, as though they were neutral instruments. As Silicon Valley entrepreneur Antonio García Martínez deceptively asserts, "The algorithm . . . is just fancy talk for a recipe of logical steps and maybe some math. Since there isn't a chance in hell our brains can parse the jumble of content . . . an algorithm sorts it out for us."[14]

Martínez's assertion of innocence aside, each of the stages of this design process deserves careful examination for its role in the introduction of bias. But to grasp the power and significance of algorithmic culture, we need to see them working in connection to produce a technology that creates a new reality whose presence we must then navigate. Reality is complex and messy, and in creating any algorithm, someone has to decide which features will be selected for inclusion in the process of development, those that will be used to train the algorithm or introduced later as an algorithm is operational. In the process of selection, features that are already assumed to "matter" and can be easily obtained, measured, datafied, and manipulated are prioritized. Further, the complexity of even those real phenomena is "transcoded,"[15] that is, reduced to what Cheney-Lippold calls a "measurable type,"[16] a sort of "ideal type" that diminishes and reshapes cultural reality into machine language that creates, constitutes, and actually causes new objects to emerge[17] as new categories of knowledge. Any concept thus "transcoded" into machine language diminishes and contributes to silencing the originary real. The loss is unavoidable, a banal effect of translation. As Michel Callon and Bruno

Latour suggest, the process will "translate" even the will of the most diligent programmer "into a language of its own."[18] In this way, in the crucial and necessary acts of selection and translation, algorithms begin their work as technologically rearticulating reality, displacing and supplanting a now degraded and largely eclipsed real world.

Beyond attention to the cultural implications of selection and translation, the implementation of algorithms has increasingly and explicitly been less about exposing truth or tendency than about being "useful," in which useful can be interpreted in a wide variety of ways. In other words, the details of the process matter less than being able to generate "useful" mechanisms for describing, operating in, and making decisions in the emerging world that serve particular ends. Being able to manipulate, visualize, use, buy, and sell big data are increasingly sufficient goals, completely apart from any truth or any match to the real world. As the algorithmic world becomes the real world, the tendency in that world is toward results not truth.

The drift toward "high tech results are truth" is exacerbated by a deeply ensconced cultural commitment to "technological progress," which as explained by Slack and Wise, "equates the development of new technology with progress"[19] and holds that technology is the ultimate "fix"[20] to any sociocultural challenge or problem. This belief and the practices that embody that belief are held to and relied on with a fervor that largely silences the ability to ask or explore what progress is and who or what benefits from it. As David Noble pointed out decades ago in an observation that is as true today as it was then, it is very nearly heresy to just ask the question, is the development of new technology *necessarily* progress?[21] Meredith Broussard calls this *technochauvinism*, "the belief that tech is always the solution."[22] Coupled with the belief that "computers get it right and humans get it wrong" and "computers are better because they are more objective than people," the emerging logic is that results *are* truth, at least the truth that matters.[23]

An example that illustrates this drift is the often-cited case of bias in hiring using algorithmic screening practices. Hiring algorithms enable handling enormous numbers of job applications rapidly; they enable bypassing the ostensible and known biases of individual human screeners; and they lead to the hiring of predictably successful employees. They are therefore considered efficient, unbiased, objective, and successful. However, in building hiring algorithms based on the attributes of previously successful employees and targeting ads based on previously successful sites for posting, these algorithms produce success at the expense of diversity, notably in gender and racial diversity, regardless of intent.[24] The algorithmic process is likely to find only what is similar to what has previously been found, thereby reinforcing old patterns of inclusion and exclusion. That bias is "baked in" to the algorithm itself has been explored by many researchers and activists.[25]

The usefulness of these algorithms is often nothing more than reproducing sameness more efficiently and obscuring the work of even more radical bias. As Frank Pasquale has put it in terms of search results, "they help create the world they claim to merely 'show' us."[26] Ruha Benjamin, in considering the racial bias in these algorithms, concludes that

> algorithms may not be just a veneer that covers historical fault lines. They also seem to be streamlining discrimination—making it easier to sift, sort, and justify why tomorrow's workforce continues to be racially stratified. Algorithmic neutrality reproduces algorithmically sustained discrimination.[27]

That these are the available worthwhile applicants creates and justifies and thereby further promotes a world in which these *are* the only available worthwhile applicants.

Critiques of bias in algorithms sometimes call for more diverse representation in the design process based on the assumption that more diverse designers will identify diverse features to be written into the algorithms. As Broussard points out, "Computer systems are proxies for the people who made them. Because there has historically been very little diversity among the people who make computer systems, there are beliefs embedded in the design and concept of technological systems that would be better off rethinking and revising."[28] However, very few people actually participate in the design teams producing algorithms, making it reasonable to expect little more than extremely limited participation by representatives who would then be expected to stand in for multiple forms of diversity. Furthermore, because those few diverse experts would likely be trained in the same largely nondiverse schools of computer and data science—taking the same courses, solving the same kinds of problems, and anticipating the same kinds of successful careers as their peers—they would be unfairly burdened by and incapable of standing in as a spokesperson and technician for all manner of diversity. These everyday realities operate against producing and fully incorporating scientists with truly innovative and diverse goals and skills.

This situation is further entrenched given the drift toward usefulness as the ultimate measure of the work once on a design team, but just as crucially given the limited opportunity to get on the team to begin with. Imagine, for example, a university setting in which a single faculty position opens up in epidemiology. The search committee has a choice between two white male candidates with expertise in the algorithms that track the likelihood of infection in the current SARS-CoV-2 epidemic and a (even if superior) diversity candidate with expertise elsewhere, perhaps in the impact on the sense of community in marginalized groups during the spread of flu viruses. One of the white guys studying SARS-CoV2 algorithms will likely win out. These kinds of real decisions get made every day and the tendency to prioritize based on momentary (if very real) conceptions of usefulness often prevail over diversity. Ultimately, however, regardless of who participates on design teams, the orientation toward usefulness operates against meaningful diversity.

Algorithms, regardless of intent, are always biased. As Cheney-Lippold elaborates, "All algorithmic interpretations produce their own corrupted truth . . . in ways particular to their technological capacity and programmed direction."[29] The process of selection sets the initial parameters for what matters and what does not, what can be seen and what is obscured, what speaks, and what is silenced. In the process of translation into computer language, the transformation creates new objects, new realities with which to describe identities and predict behaviors constructed in the process itself. That this contains bias is unavoidable. That we tend to ignore those biases is not inevitable, but is conditioned by deep cultural commitments to technological progress and an emerging understanding of and commitment to usefulness as the measure of a successful algorithm. No single design team can counter the hegemonic power of those forces.

ALGORITHMS IN CONTROL: MACHINE LEARNING

We've never talked with a data scientist who can explain what actually happens in "machine learning." They always resort to some manner of "hand waving" accompanied by the statement that "we don't actually know what happens." For the data scientist, machine learning means the capacity of an algorithm "to adapt to new circumstances and to detect and extrapolate patterns."[30] This "learning by experience" takes different forms in supervised, semi-supervised and unsupervised learning. As described by Cheney-Lippold, supervised and semi-supervised refer to initial human intervention to "draw the initial limits for categorical identity," which means that the programmer presets the identities on which the algorithm operates. For example, "man" would be coded with a specific list of attributes that would not change as the algorithm operates using that coded identity. Unsupervised or unstructured learning "can be thought of as finding patterns in the data above and beyond what would be considered pure unstructured noise," letting "new categorical limits arise based on data's statistical proximity." In unsupervised learning, an identity such as "man" isn't relevant as a fixed identity, but might merely be the identifier used to point to emergent correlations and patterns in the data.[31] These correlations "actually create and re-create kinds of people in the process of naming and studying, which becomes a materialization of the scientific imagination" that may have little or no resemblance to what anyone thinks a man is.[32]

In turning an algorithm lose to find and create previously unimagined patterns, data scientists don't really know what happens computationally to make those patterns emerge. The algorithms are too complex, and the computational work is too fast, both far beyond human time and understanding. To ask what is happening is like asking of human cognition, "What is thinking?" We know it happens and we can point to outcomes, even often consistent outcomes, but we don't know for certain what is happening in its production. Data scientists are in a similar position. It isn't clear that they even care, because they are more interested in and rewarded for consistent and useful results than for considering the cultural implications of an essentially unknowable process. All that matters is whether or not the mac 'n cheese is edible, or whether or not you have identified a criminal.

In spite of the necessary limitations of understanding machine learning technically or the difficulty translating the math into nontechnical language, it is important to acknowledge the role of machine learning as a precipitate in algorithmic culture, for the consequences are considerable. Probably the best mathematically accessible example of the kinds of consequences is Broussard's extended exploration of the causes of mortality in the sinking of the Titanic.[33] Broussard illustrates how a supervised machine learning process with known, selected, and mathematically translated data about the passengers on the Titanic produces a pattern that with 97 percent accuracy learns "that passenger fare is the most important factor in determining whether a passenger survived the Titanic disaster,"[34] with those who paid more having higher survival rates. Broussard cautions, however:

> It would be unwise to conclude from this data that people who pay more have a greater chance of surviving a maritime disaster. Nevertheless, a corporate executive could easily argue that it would be statistically legitimate to conclude this. If we were calculating insurance rates, we could say that people who pay higher ticket prices are less likely to die in iceberg accidents and thus represent a lower risk of early payout. This would allow us to charge rich people less for insurance.[35]

Broussard counters the algorithmic result with what is considered anecdotal information that might better account for the differences in survival rates: elements of social context that might more truthfully account for the differential rates of survival. For example, she draws on correspondence that suggests that what mattered was how one jumped off the ship. This "data" was not part of the original set, but rather became evident through qualitative contextual reading. She notes that such unscientific information may not be readily available or conducive to coding, because as we know from the above discussion on selection, "Not everything that counts in counted. The computer can't reach out and find out the extra information that might matter."[36] She concludes that "part of the reason we run into problems when making social decisions with machine learning is that the numbers camouflage important social context."[37]

Yet we do increasingly make significant social decisions based on machine learning without human intervention and assessment, and we do so without concern for important contextual elements that are actively camouflaged. Besides the underpinning of an unexamined commitment to technological progress, the overall shift toward automation to eliminate human labor (which began with the "industrial revolution") and thereby improve efficiency and maximize profit now drives the shift to rely on algorithms to manage production and make significant social decisions, thereby reshaping culture. As the era of cheap labor wanes,[38] and every conceivable industry and service searches for ways to replace that labor with algorithmically automated processes, the so-called "fourth industrial revolution" will run on algorithms. Klaus Schwab's version of this story is just one of the many available maps pointing to where we are headed in this "bright future" and how we should respond.[39] Bucked up by a sense of misplaced inevitably, Schwab, like generations of technobuskers before him, urges us to "step right up":

> Technology and digitization will revolutionize everything. . . . Simply put, major technological innovations are on the brink of fueling momentous change throughout the world—inevitably so. . . . Digitization means automation, which means that companies do not incur diminishing returns to scale (or less of them, at least). . . . The fact that a unit of wealth is created today with much fewer workers compared with 10 or 15 years ago is possible because digital businesses have marginal costs that tend toward zero.[40]

Widely circulated articles and online videos speculate about the jobs that will disappear due to algorithmically driven automation, thereby instructing people to adjust their lives in anticipation of a dramatically transformed world. Those disappearing jobs include drivers, farmers, printers and publishers, cashiers, travel agencies, manufacturing workers, dispatchers, waiting tables and bartenders, bank tellers, military pilots and soldiers, fast-food workers, telemarketers, accountants and tax preparers, stock traders, construction workers, and movie stars.[41] *Business Insider* has an expanded list of thirty-seven disappearing jobs that add to these positions such as postal service workers, sewing machine operators, textile workers, secretaries and administrative assistants, switchboard operators, miners, computer operators, mechanics, and all manner of machine operators and technicians.[42] There are so many jobs on these lists that it is difficult to imagine what anyone will do in this brave new world to earn a living. Anxiety about the shrinking opportunities for rewarding work in this algorithmic world is inevitable.

NEGOTIATING POWER AND CONTROL: ALGORITHMS AND USERS

One of the most elusive elements of algorithmic culture is the way in which humans interacting with algorithms marks a change in work life and practices and perceptions of autonomy and control. In short, we are experiencing a shift in the distribution of power and control in human-machine relationships. Slack and Wise insist that when thinking about technology, we recognize that technology is not a thing, but an "articulation," that is, "*a contingent connection of different elements that, when connected in a particular way, form a specific unity.*"[43] Those elements include "*physical arrangements of matter, typically labeled technologies, and a range of contingently related practices, representations, experiences, and affects.*"[44] The laborer or user of something designated as technology is as much a part of that contingent unity as is the way in which control is distributed among those complex articulations.[45] For example, the so-called user of a smart speaker such as Alexa or Echo is integral to the technology, not strictly speaking a user of it, in that the thing only "works" or "does its thing" in relation to the actions and activities of the "user." It is the relationship that allows for certain voices and actions in certain circumstances (plugged in, connected to the internet, set at a prescribed distance, etc.) to have certain effects.

Automation almost always involves, usually intentionally, a concomitant process of deskilling of the workforce, that is, removing the power of decision making from the workers. The value in doing so for management is clear:

> A knowledgeable, decision-making skilled worker is never fully under management's control. Therefore, it is in management's interests to learn the worker's skills, train others in those skills, or better yet, create a machine to replicate those skills.[46]

Totally apart from whatever the mix of motives—reducing the cost of production, speeding up or increasing production, eliminating problems of absenteeism or a lack of available workers, creating a consistently standard product, enhancing safety, or just refusing to negotiate with skilled labor—the deskilling of labor with the implementation of algorithmically controlled production gives management an upper hand over labor. It also means that decisions and the effects of those decisions are increasingly influenced by the outcomes of machine learning. And remember, nobody really knows what happens—even mathematically—in the machine learning process, so they certainly don't have a sense of the ways that biases might enter through the process. Further, recall that the outcomes of machine learning are part of a process of creating new realities that may or may not make a good match with reality as most of us see or experience it. The consequences can be enormous. Remember, too, that it isn't the algorithm's "fault" anymore that Alexa is at "fault" alone for—or in control of—ordering that unwanted book. Rather, once again, the devil is in the details of the articulation of many factors.

An excellent contemporary example that illustrates the shift in the distribution of power and control through algorithmically induced deskilling and what can go wrong when that happens is the case of the two deadly crashes of *Boeing* 737 Max aircrafts on October 29, 2018 and March 10, 2019. Following the developing accounts of these crashes has been fascinating and instructive, especially in terms of the relationship between automation, algorithms, and the way users interact with algorithms. Although it is certain we will never have a complete picture of the interrelationships among the wide variety of elements that articulated to result in these crashes—and there are multiple versions of explanation and blame—we draw our discussion from pilot and respected investigative journalist William Langewiesche's coverage of the question, "What Really Brought Down the Boeing 732 Max?"[47] His account benefits from technical familiarity and a passion to explore the complexity of the problem beyond telling a good story.

These crashes have become almost legendary, dominated by the breathtaking and terrifying image of a computer algorithm pushing the nose of the aircraft down for short bursts and letting up for a bit and repeating this over and over while the pilot and copilot fight with manual controls to override the computer and race through the flight manual trying to find directions for how to respond until the craft is wildly thrown out of trim and crashes nose down into the sea in one case and into the ground in the other. Langewiesche summarizes the crashes in this way:

> After both accidents, the flight-data recordings indicated that the immediate culprit was a sensor failure tied to a new and obscure control function that was unique to the 737 Max: the Maneuvering Characteristics Augmentation System (MCAS) [aka, an algorithmically driven system]. The system automatically applies double-speed impulses of nose-down trim, but only under circumstances so narrow that no regular airline pilot will ever experience its activation—unless a sensor fails. Boeing believed the system to be so innocuous, even if it malfunctioned, that the company did not inform pilots of its existence or include a description of it in the airplane's flight manuals.[48]

Langewiesche also summarizes the dominant narrative that emerged shortly after the accidents:

> [Boeing] had developed the system to elude regulators; that it was all about shortcuts and greed; that it had cynically gambled with the lives of the flying public; that the Lion Air pilots were overwhelmed by the failures of a hidden system they could not reasonably have been expected to resist; and the design of the MCAS was unquestionably the cause of the accident.

In other words, because the pilots were unaware of and/or incapable of countering the actions put in place by the algorithm, the MCAS caused the crash.

Langewiesche, in contrast to assigning blame to the algorithm, rather mercilessly attributes the accidents to the pilots:

> What we had in the two downed airplanes was a textbook failure of airmanship. In broad daylight, these pilots couldn't decipher a variant of a simply runaway trim and they ended up flying too fast at low altitude, neglecting to throttle back and leading their passengers over an aerodynamic edge into oblivion.[49]

Our point is not to assign blame one way or the other, but to notice, even in Langewiesche's own writing, there is never so simple an assignment of blame. Certainly, it is not simply the algorithm at fault: the pilots lacked "airmanship"; they "neglected" to perform appropriately, even in the second crash after "they had been briefed on the MCAS system and knew the basics." Equally, however, it is never just the pilots at fault: Langewiesche acknowledges "the MCAS as it was designed and implemented was a big mistake"; the pilots "couldn't decipher a variant" of an otherwise normal occurrence, and directions for responding were not in the flight manual.[50]

A far better way to understand the loss of these hundreds of lives is to acknowledge that something went very wrong in the relationship between the algorithm and the user, placing a far greater burden on the user than was normal for those pilots. Automated features on aircraft are designed to minimize pilot error and minimize accidents, and they typically do so "as long as conditions are routine," as Langewiesche puts it. But as tasks are computerized and routinized, they contribute to deskilling the user. Moments of crisis require "airmanship," a "visceral sense of navigation, an operational understanding . . . the ability to form mental maps . . . fluency in the nuance . . . a deep appreciation for the interplay between energy, inertia, and wings."[51] If you never fly a plane manually, if you are never routinely called to respond to novel situations, you will never have the opportunity to develop those skills, the exact skills called upon to negotiate with the algorithm, to resist it as necessary, whether you know it is there or not.

The story of the 737 Max crashes involves more than an algorithm and a couple of pilots. That original narrative of blame shared by Langewiesche should not be too quickly dismissed, for even in his own telling, once again, there *are* companies in competition with one another (*Boeing* versus *Airbus*); companies that do cut corners to beat out another, corners that in a different culture of responsibility and a different economy they would not have cut. There *are* in the story airlines that not only cut corners but cheat on their maintenance records and certify pilots whose capabilities are questionable. There *are* politics that enter into what can be investigated and said about these crises. There is never just a guilty algorithm or a guilty pilot in these stories. There is, however, an algorithmic culture that, for many reasons, both good and bad, enables someone like Langewiesche to attribute control and shift the blame to the users, the ones less able to defend themselves, the ones who are now expected to develop airmanship in circumstances and structures that operate against attaining those very same skills. That is also the work of an algorithmic culture: it empowers the algorithms and those who employ them and disempowers the users of those algorithms, and places new excessive demands on them.

We are all increasingly in that position in algorithmic culture: of having to learn on our own how to negotiate with and manage the algorithms that are indeed largely hidden and that proliferate with the backing of powerful companies, governments, economies, and inequitable systems of education, healthcare, and justice. Yet we are individually held responsible for any failures because the assumption remains that success and failure are choices entirely within our individual control, even as the opportunities for taking control are taken from us, even as the algorithms get smarter at the expense of the "street smarts" we might collectively call on to thrive in the face of the growing crises that characterize contemporary life.

ALGORITHMIC CULTURE: THE PRODUCTION OF COMPUTATIONAL LOGICS

Machine learning algorithms increasingly make decisions that we expect to be made by humans, decisions that allow "officials to outsource decisions that are (or should be) the purview of democratic oversight."[52] In the process of designing and implementing those algorithms, new versions of the world are created and reinforced that then become the ground on which further (often algorithmic) decisions are made. To matter in this emergent world, in the "new real," we are encouraged to become better data selves: to define ourselves and shape our lives to conform to and satisfy the parameters of the algorithms. So, we do things as mundane as shape job applications to include the keywords that we know will get our résumés considered at a higher level of algorithmic consideration. We purchase a house, not because we want one, but because it will improve our algorithmically generated credit rating. In school, we become more concerned about grade point than learning, because algorithms "understand" grade points but not the immeasurable wisdom learning might bring. We reduce education to computational learning outcomes for exactly the same reason. We read the books and watch the movies algorithms teach us we like. We buy the things we are supposed to. We believe in the stereotypes that algorithmically determined content feeds us over and over again. We make decisions about incarceration, parole, and the tracking of students in school based on racist and sexist algorithms. We even transform spiritual activities, like running or meditating, by measuring success in terms of minutes, steps, and heartbeats rather than letting them be private, personal, and unknowably good. All that is measurable is elevated and transformed. All that is unmeasured is trivialized, unnecessary, indulgent luxury, or the wasteful activity of marginalized people. In responding appropriately in the presence of algorithms, we produce algorithmic culture without thinking, without even knowing we are doing so.

"If you can't measure it, it isn't real," a sociologist said in 1975. It was funny at the time (and a few people dropped his class). Who would have imagined he would be so right, the only difference being that his real is the "new real," the one he helped create, the one we might have avoided if people like him had taught differently. Things can, after all, be different.

We began this chapter emphasizing our affinity with Raymond Williams's concept of the structure of feeling. We wrote that "while such a structure is not fixed, uniform, or permanent, it is a set of relationships that exhibits some tenacity that has consequences." We are indeed undergoing a moment of profound change in how we interact with the earth, in how we are governed, and in relations of privacy, sociality, race, criminality, justice, and freedom. Most of all we are undergoing an enormous transformation in what it means to be human, to be a self, even just to be. We are not two: a self and a data self. We are complex beings caught up in a web of sometimes homologous, sometimes contradictory forces, values, beliefs, and practices that contribute to an affective sense of chaos and confusion. What is real? What is the "new real"? How and where are we bits of both at the same time? While we keep trying to draw lines between the two, as between real news and fake news, privacy and sharing, independence and dependence, we acquiesce, struggle, and resist, in spite of ourselves, already in the presence of algorithms that constrain and enable the choices we make. As this structure of feeling emerges, it marginalizes, represses, silences, eclipses, and disappears what cannot be counted and what can be counted but counted out.

To the extent that silencing is successful, nobody would drop a class because the professor insisted that the real is only that which can be counted. To the extent the silencing is successful and algorithmic culture becomes hegemonic, we are caught up in a loop where selection is based on what can be datafied and deemed useful, where what is datafied produces a new real outside human judgment or democratic processes, where users negotiate that real in processes that reproduce and augment sameness and long-standing patterns of discrimination, and where power and control reside in the hands of those who produce and employ the algorithms at the expense of users as workers. Without challenging this circularity, algorithmic culture will continue to silence difference, growth, creativity, and change. This is why we need the concept of algorithmic culture: to *see* its often-hidden operation, to *locate the connections* where the work is performed and where intervention might matter, and to *challenge and resist* its ascension in ways that matter.

NOTES

1. Snowden, Edward. "Edward Snowden on Trump, Privacy, and Threats to Democracy." *The 11th Hour. MSNBC* (September 17, 2019).
2. Ibid.
3. Williams, Raymond. *The Long Revolution*. London: Chatto & Windus (1961), 63.
4. Williams, Raymond. *Marxism and Literature*. Oxford: Oxford University Press (1977), 131.
5. Williams, Raymond. *Television: Technology and Cultural Form*. New York: Schocken Books (1975), 26.
6. Ibid., 27.
7. Slack, Jennifer Daryl and J. Macgregor Wise. *Culture and Technology: A Primer*, 2nd edition. New York: Peter Lang (2015), 153.
8. Galloway, Alexander R. *Gaming: Essays on Algorithmic Culture*. Minneapolis, MN: University of Minnesota Press (2016).
9. Striphas, Ted. "Algorithmic Culture." *European Journal of Cultural Studies* 18(4–5) (2015), 396.
10. Ibid., 408.
11. Ibid., 396–397.
12. Ibid., 404.
13. Russell, Stuart J. and Peter Norvig. *Artificial Intelligence: A Modern Approach*, 3rd edition. Uttar Pradesh, India: Pearson (2015), 121.
14. Martínez, Antonio García. *Chaos Monkeys: Obscene Fortune and Random Failure in Silicon Valley*. New York: Harper Collins (2016), 506.
15. Manovich, Lev. *The Language of New Media*. Cambridge, MA/London: MIT Press (2001), 45.

16. Cheney-Lippold, John. *We Are Data: Algorithms and the Making of Our Digital Selves*. New York: New York University Press (2017), 47.

17. Ibid., 46.

18. Callon, Michel and Bruno Latour. "Unscrewing the big Leviathan: how actors macro-structure reality and how sociologists help them do so." In *Advances in Social Theory and Methodology: Toward an Integration of Micro- and Macro-sociology*, edited by K. Knorr-Cetina and A. V. Cicourel. Boston, MA/London/Henley: Routledge and Kegan Paul (1981), 277–303.

19. Slack and Wise *Culture and Technology*, 28.

20. Ibid., 153.

21. Noble, David. "Introduction." In *Architect or Bee? The Human/Technology Relationship*, edited by Shirley Cooley. Boston, MA: South End Press (1982), xi–xxi.

22. Broussard, Meredith. *Artificial Unintelligence: How Computers Misunderstand the World*. Cambridge, MA: The MIT Press (2018), 7–8.

23. Ibid., 8.

24. Bogen, Miranda. "All the ways hiring algorithms can introduce bias." *Harvard Business Review* (May 6, 2019).

25. O'Neil, Cathy. *Weapons of Math Destruction*. Crown, 2016; Eubanks, Virginia. *Automating Inequality: How High-Tech Tools Profile, Police, and Punish the Poor*. Boston, MA: St. Martin's Press (2018); Cheney-Lippold, John, *We Are Data: Algorithms and the Making of Our Digital Selves*. New York: New York University Press (2017); Ruha, Benjamin. *Race after Technology*. Cambridge, UK/Malden, MA: Polity Press (2019); Pfefferkorn, Marika. Coalition to Stop the Cradle to Prison Algorithm Celebrates Hard-Won Victory with the Dissolution of Problematic Data-Sharing Agreement." *Dignity in Schools* (January 29, 2019).

26. Pasquale, Frank. *Black Box Society: The Secret Algorithms that Control Money and Information*. Cambridge: Harvard University Press (2015), 61.

27. Benjamin, "Race," 143.

28. Broussard, *Artificial Unintelligence*, 67.

29. Cheney-Lippold, *We Are*, 12.

30. Russell and Norvig, *Artificial Intelligence*, 3.

31. Cheney-Lippold, *We are*, 79.

32. Benjamin, "Race," 117.

33. Broussard, *Artificial Unintelligence*, 96–119.

34. Ibid., 110.

35. Ibid., 114.

36. Ibid., 116.

37. Ibid., 115.

38. Patel, Raj. *A History of the World in Seven Cheap Things: A Guide to Capitalism, Nature and the Future of the Planet*. University of California Press (2018).

39. Schwab, Klaus. *The Fourth Industrial Revolution*. New York: Currency (2016).

40. Ibid., 9–10.

41. Alux.com. 2018. "15 jobs that will disappear in the next 20 years due to AI." https://www.alux.com/jobs-gone-automation-ai/. Accessed June 14, 2020.

42. Gillett, Rachel, Andy Kiersz, and Ivan De Luce. "37 Jobs That Could be Decimated by 2026." *Business Insider* (July 16, 2019).

43. Slack and Wise, *Culture and Technology*, 152.

44. Ibid., 153.

45. Ibid., 59–73.

46. Ibid., 64.

47. Langewiesche, William. "What Really Brought Down the Boeing 737 Max?" *The New York Times Magazine* (September 18, 2019).

48. Ibid.

49. Ibid.

50. Ibid.

51. Ibid.

52. Benjamin, "Race," Technology. Cambridge, Polity Press, 2019, 53.

BIBLIOGRAPHY

Alux.com. 2018. "15 jobs that will disappear in the next 20 years due to AI." https://www.alux.com/jobs-gone-automation-ai/. Accessed June 14, 2020.

Benjamin, Ruha. *Race after Technology: Abolitionist Tools for the New Jim Code*. Cambridge, UK/Malden, MA: Polity, 2019.

Bogen, Miranda. 2019. "All the ways hiring algorithms can introduce bias." *Harvard Business Review*. https://hbr.org/2019/05/all-the-ways-hiring-algorithms-can-introduce-bias. Accessed June 11, 2020.

Broussard, Meredith. 2018. *Artificial Unintelligence: How Computers Misunderstand the World*. Cambridge, MA: MIT Press.

Callon, Michel and Bruno Latour. 1981. "Unscrewing the big Leviathan: how actors macro-structure reality and how sociologists help them do so." In *Advances in Social Theory and Methodology: Toward an Integration of Micro- and Macro-sociology*, edited by K. Knorr-Cetina and A. V. Cicoure, pp. 277–303. Boston, MA/London/Henley: Routledge and Kegan Paul.

Cheney-Lippold, John. 2017. *We Are Data: Algorithms and the Making of Our Digital Selves*. New York: New York University Press.

Eubanks, Virginia. 2017. *Automating Inequality; How High-Tech Tools Profile, Police, and Punish the Poor*. New York: St. Martins.

Galloway, Alexander R. 2006. *Gaming: Essays on Algorithmic Culture*. Minneapolis: University of Minnesota Press.

Gillett, Rachel, Andy Kiersz, and Ivan De Luce. 2019. "37 jobs that could be decimated by 2026." *Business Insider*. https://www.businessinsider.com/jobs-quickly-disappearing-in-the-us-2017-5. Accessed June 14, 2020.

Langewiesche, William. 2019. "What really brought down the Boeing 737 Max?" *The New York Times Magazine*. https://www.nytimes.com2019/09/18/magazine/boeing-737-max-crashes.html. Accessed September 19, 2019.

Manovich, Lev. 2001. *The Language of New Media*. Cambridge, MA/London: MIT Press.

Martínez, Antonio García. 2016. *Chaos Monkeys: Obscene Fortune and Random Failure in Silicon Valley*. New York: Harper Collins.

Noble, David. 1982. "Introduction." In *Architect or Bee? The Human/Technology Relationship*, edited by Shirley Cooley, pp. xi–xxi. Boston, MA: South End Press.

O'Neil, Cathy. 2017. *Weapons of Math Destruction: How Big Data Increases Inequality and Threatens Democracy*. New York: Broadway Books.

Patel, Raj. 2018. *A History of the World in Seven Cheap Things: A Guide to Capitalism, Nature and the Future of the Planet*. Oakland: University of California Press.

Pfefferkorn, Marika. 2019. "Coalition to stop the cradle to prison algorithm celebrates hard-won victory with the dissolution of problematic data-sharing agreement." *Dignity in Schools*. https://dignityinschools.org/coalition-to-stop-the-cradle-to-prison-algorithm-celebrates-hard-won-victory-with-the-dissolution-of-problematic-data-sharing-agreement/. Accessed June 11, 2020.

Pasquale, Frank. 2015. *The Black Box Society: The Secret Algorithms That Control Money and Information*. Cambridge, MA: Harvard University Press.

Russell, Stuart J. and Peter Norvig. 2015. *Artificial Intelligence: A Modern Approach*, 3rd edition. Uttar Pradesh, India: Pearson.

Schwab, Klaus. 2016. *The Fourth Industrial Revolution*. New York: Currency.

Slack, Jennifer Daryl and J. Macgregor Wise. 2015. *Culture and Technology: A Primer*, 2nd edition. New York: Peter Lang.

Snowden, Edward. 2019. "Edward Snowden on Trump, privacy, and threats to democracy." *The 11th Hour. MSNBC*. https://youtu.be/e9yK1QndJSM. Accessed 05/29/2020.

Striphas, Ted. 2015. "Algorithmic culture." *European Journal of Cultural Studies*, 18(4–5), 395–412.

Williams, Raymond. 1961. *The Long Revolution*. London, Chatto & Windus.

———. 1975. *Television: Technology and Cultural Form*. New York: Schocken Books.

———. 1977. *Marxism and Literature*. Oxford: Oxford University Press.

Chapter 2

Fetishizing Algorithms and Rearticulating Consumption

Soonkwan Hong

Big data is one of the most iconic networks in human history. It mushrooms all domains of our lives, and the corollary has been evident long enough for individuals and organizations to make sense and use of the technological esoterica.[1] The emergence of big data has conjured up another immaterial object, namely, artificial intelligence (AI), which may not be called an entity because it has never been fully embodied or realized.

It is natural to be still unfamiliar with such human creations, but history has seen the likes since Hephaestus' inventions and Homeric automata. They have existed long enough in myths and in imaginaries but not as an authenticated being in our minds. Perhaps we do not want to or simply cannot imagine the future, as Jameson judiciously admonishes.[2]

The lack of imagination has created the most marketable image, idea(l), concept, and mode of being: the future. Books, movies, products, and visions all enjoy a fair share of consumerism that recycles and ultimately perpetuates the original version of future, which only signifies difference and change without many specifics. Because the mantra for success in the market is no different from what the future may hold and provide, consumers have been hypnotized by the potentially spurious superiority of what has been nonexistent. Innovative technology (i.e., big data and AI) may be crucial to sustain the market economy, but it can also be encumbering to reach the very future humanity has long projected. Amid this paradox, consumers are baffled at best with the future that appears to desperately need nonhuman agents to sustain humanity. At this critical juncture in late modernity, no discourse can afford to forego the transformative entanglement between sociocultural imagination and the political economy of technology.

In this chapter, therefore, I offer a critique of the current state and nebulous politics of the new sociotechnical environment, relevant practices, and politics of algorithms vis-à-vis consumer culture. Such a discussion will help analyze the nexus between algorithms and consumption, based on some observations in the current consumer culture that necessitate a transformation of consumer subject. To provide more lucid descriptions and substantive discussions, followed by some considerations for future scholarship in this highly "contagious" field of study, I scrutinize algorithms as machines, networks, and nonhuman actors in the market that encompass the internet of things, AI, big data, and all other forms and functions of the fetishized assemblage.

FETISH FOR ALGORITHMS

Consumers' lived experiences have been sequestered by algorithms that inherently induce individuals to fetishize privacy, security, convenience, productivity, accuracy, and control. The notion of sequestration is pivotal in discussing the universal fetishization wherein institutionalizing motives and practices reconfigured sociopolitical relations, individual experiences, time, and space.[3] Structuration theory informs that structures enable and emancipate—as well as prescribe and oppress—the populace.[4] However, purely natural and highly personal domains, stages, and aspects of life can be the easiest targets of sequestration. Consumption since modernity has stayed in the most natural, real, and personal realm of all human practices. Due to the sequestration of consumer experiences, consumption has become paradoxically punctual and precise in terms of time, location, amount, frequency, and even the level of pleasure. These new features of consumption in algorithmic consumer culture formulate new market(able) values and systemic priorities that stimulate secondary fetishization, which builds more palpable pressure on the ever-threatened human relationships in the market as a modern institution.

Algorithms reward and punish consumers based on the constant negotiation (between consumers and algorithms) for the quality of friendship, collegiality, and even love. Social media is the all-time great platform to isolate individuals in the grid of deep connectedness, which should promote more meaningful human(e) relationships. Emotions are stripped, compassion is up for sale, and transparency is a sardonic luxury. Voluntary "lifelogging" starts as early as kindergarten, using various media on mobile and wearable devices, which fuel the algorithmic engine that crosses all domains of life. It is all possible because sequestration provides pseudo-agency, quasi-privacy, and a simulated feeling of inclusion for consumers. Together, these entice consumers to further fetishize algorithms. An algorithmic social away from *the social* is a premeditated milieu for predetermined behaviors and discourses that perpetuate the sequestration-fetishization cycle. The predictability of the social and the subsequent prêt-à-porter psyche for all reduces the social to a mere approximation of what used to be grand, complex, and yet highly intimate.

Connectedness in current times even reaches the sacred territory of humanity, the body. Our bodies are sequestered as we delegate our rights and abilities to develop, maintain, and control the body to algorithms by allowing them to problematize our bodies and minds. We find ourselves no longer active without step-counting wristbands and less competent and productive without notifications. We no longer have to confront our own corporal indolence and cerebral limitations because they are taken from us and returned in a form of body techniques. Consumers must learn these new techniques to stay in the game. For Marcel Mauss, body techniques are the regimented and almost uniform ways of life and sociocultural practices that embody the principles and superficial elements of a given culture.[5] The techniques that algorithms utilize are collectively "crafted" and subsequently learned by the mass. This collective aspect of body techniques inexorably leads to shaping, maintaining, and reshaping rituals of the culture.[6] Algorithmic bodily rituals are deeply entrenched in the current consumer culture, ranging from the "Breathe" feature on *Apple* watches to taking hyper-personalized multivitamin tablets algorithmically cocktailed by a company, *Baze*, based on a small amount of the consumer's blood. When the body is sequestered, the political landscape around it is also dismantled.

Biopolitics operate by controlling the population beyond norms, ethics, and laws and by invalidating class struggle through mandatory measures over the entire population.[7] Biopolitics also facilitates "governing the body" (alive or dead) to become a purely public and political agenda through vaccination, birth control, public hygiene education, and infertility treatment legislation, to list a few.[8] Such practices and processes do not discriminate social classes and, therefore, they obligate new ways of governing all bodies. Algorithms play a critical role in the makings and workings of biopolitics by providing "accurate" figures and statistics. Quantification of the body, however, is not algorithms' sole contribution to biopolitics. Individuals are made to believe they have full control of the body because algorithms share a supposedly fair amount of information with the actual owners of the body through doctor visit reports and healthcare applications provided by hospitals. The rest of the information stays in medical algorithms, such as the Massachusetts General Hospital Utility Multi-Programming System (MUMPS), Apgar to evaluate a newborn's condition at birth, and APACHE to determine the severity of patients in intensive care units.

The notion of neoliberal governmentality as the basis and preliminary state of biopolitics explains how algorithms can transform individuals into self-governing subjects with numerous choices.[9] These sociopolitical practices together encourage the population to adopt, of their own free will, new body techniques for their lives—but only within methodically calculated boundaries. Institutions and governments, with the aid of algorithms, domesticate individuals with prescribed techniques of governing the body in almost all stages and aspects of life.[10] Evidently, consumers have started fetishizing algorithms for other values they promise to provide—health and longevity.

Algorithmic fetishization has taken us full circle to return to the original meaning of *fetish.* Algorithms as objects are impregnated with specious abilities that are not originally part of their teleological account. Fetishized algorithms extend and intensify their reach in all domains of society; as a result, consumers tend to believe in algorithms' contributions to all cultural and economic outcomes, which, in fact, is often nominal or even negative.[11] The repeated fetishization brings another delusion to consumers in which they misconstrue that the sum of positives is greater than the whole negativity from algorithms. Dystopian possibility is casually negated, while grand narrative is still in operation in a different tone and manner (i.e., the Fourth Industrial Revolution) that again raves about enlightenment and progress through science and technology.

CHOREOGRAPHED CONSUMPTION: ACQUIESCENCE TO ENGINEERED MANIPULABILITY

Despite the stealth fetishization project that has long been in place in the market, consumer research has shown a somewhat reserved stance to the algorithm laden transformations most consumers experience daily. In the literature, there have been some discussions on a few specific features and outcomes of the data-intensive sociocultural innovation: self-quantification,[12] sentiment analysis using big data,[13] and recommendation mechanisms.[14] Among recent research on algorithmic consumption, John A. Deighton discusses the unique history, essential characteristics, and potential perils of big data.[15] His analysis and arguments inspire the field of consumer culture to redraw the big picture of algorithms as something more than a tool, a strategy, a manifestation of social interactions and relationships, or just a giant machine as a whole. Algorithms connote much more than we can individually and collectively decipher, just as does consumption.

We desire and consume symbols in the forms of material objects and intangible services because those symbols can be signified with individuated meanings as consumed and subsequently interpreted. Consumer culture has emerged as a product of the consumers' negotiation process dealing with various cultural conditions. In the process, consumers have developed an ability to signify navigating through the cultural environment where marketers constantly co-opt consumers' lived experiences and presentations of values in the lifeworld. The ability to signify—because we all possess it with varying degrees—is the self-efficacy to value signs and individuated meanings more than practical, marketized, and socially constructed symbols in the market and to (re)create and practice idiosyncratic identity through signification. What algorithms in the market do is make symbols and signs interchangeable, which confuses many consumers about their values and identities.

Leach explains that a *symbol* typically represents an entity as a whole, not as a part of the entity, but a *sign* is only accountable for a fraction of the entity.[16] Consumer identity projects are not necessarily limited to a single and invariable identity but advocate a transformative, adaptive, and essentially reflexive identity.[17] Algorithms meddle with the process by suggesting and even imposing symbolized offerings as though they are singularized.[18] Personalization based on algorithmic engineering of need and want is synonymous with exploitation of consumers' "arbitrariness" and imagination in their identity work through consumption. Whereas Barthes[19] and Saussure[20] concur that the arbitrariness is indispensable to signification, scientists and engineers do not design algorithms to tolerate anything random, unpredictable, and/or uncategorizable with a high level of peculiarity and originality.

Algorithms crush semiotics in the market. The semiotic triangle is no longer relevant in the production-consumption scene. Algorithms produce symbols using statistics and mathematical models, which are extensively used in many products and services (referent), and the interpretations are almost uniform. In other words, brands and products must become icons with an algorithmic certification to be successful in the market. The only relationship that is marketable and sustainable is based on symbol grounding between the object (i.e., products and brands) and what it symbolizes. The relationship between consumers and symbols is so often manufactured by algorithms that the social relations based on social grounding are no longer the property of humans but of algorithms. Because algorithms appropriate both symbol and social groundings, the private grounding that occurs between the object and the consumer becomes devoid of meaning and history. These should be critical elements for social relations and indispensable ingredients for signification.

Iconic brands were once developed and experienced in sociocultural relations in the market, but now icons are the product of mentions, searches, tweets, pop-ups, and reviews that constantly optimize the consumer experience and even the level of satisfaction from the very algorithmic consumption. Because what a *Tesla* Cybertruck means to any consumer is already prescribed, consumers have only a binary choice: thumbs up or down. It can be argued that marketers (brand managers to be more precise) have long used the exact practice (prescription of meanings), but consumers used to have higher negotiation power (e.g., Giesler 2008). Sociocultural compromises and endless negotiations between consumers and brands are the quintessence of a consumer identity project that can be completed only in the market. The loss of such practices ultimately cripples consumption as a means to achieve and maintain plurality.

The cultural turn in the modern market that has been enjoyed for a few decades is again on the brink of becoming one way, hyper-structured, rigid, gendered, and stultified. The conformity orientation of algorithms supports only new consumer subjects who are compatible with all aesthetic, ethical, and ideological claims by algorithms. The intention here is not to bring the agency-structure debate back to the table, but to question the nature of the structure we, as consumers, will confront for an indefinite period.[21] Intrinsically, the new algorithmic structure is not sophisticated enough to encourage consumers' re-interpretation, modification, contextualization, or re-creation of cultural meanings from marketized texts, objects, ideas, and lifestyles. Although those agential practices are still possible, algorithms always co-opt and "re-engineer" the outcome to regain control and remain adaptable but standard.

Algorithms disclose almost too much information to consumers in the name of transparency and fairness, but they also obscure the fact that much more could be hidden. What must be hidden at all times is the essence of our identities as consumers in the market. Identity had always oscillated between one that was assigned and another that was asserted. A constant negotiation was necessary to respond to a haunting yet fundamental question: Who am I in this market society? Consumers' reflexive response to the question was encouraged and expected, but algorithms now answer the question through Siri, Cortana, and Alexa for our choices of music, movies, products, brands, vacation destinations, food, and of course drugs, to make just a shortlist. The choice of what to consume is always interwoven with the choice of identity, but when the state of power relations in the market changes, the choice becomes oxymoronic. An assigned identity does not leave any room for the consumer to "embellish" her identity through consumption. At the same time, a totally asserted identity yields a rejection from the observer or causes a misunderstanding of the identity. Amid a market transformation by algorithms, a balance—or at least the chance to negotiate—between the two types of identity is lost. Technical homogenization is the underlying theme of algorithmic consumer culture, and multiplicity of identities is a systematic glitch that requires algorithms to perform continuous cultural "debugging" through location services, trending, and all kinds of subscription boxes.

(UN)PLUGGED?

Algorithm is Matrix. It is a meta-system that appears to be a compact and swift vehicle of ideas that needs only a password to be activated. Nevertheless, it is still an object; yet, it speaks to us and to one another. It is an object unavailable for consumers in its original form, which is a sociotechnical reservoir of unchartered designations and assimilations. It is also a language we will never learn to speak, but we are at least becoming inured to the sheer spread of the language with or without conscious adoption. Having to realize that the system is always there—and so we are a part of the system—is the very inconvenient truth as a temporal and variable presentation of the *Truth*.

The system (algorithms) does not embrace a breakout by a consumer; it registers it as ontological madness or anomaly at best. The system can address all other anomalies because consumers have joined the system as subsystems with or without their consent. Many consumers remain "plugged" because the system provides ontological security (albeit superficial), and "unplugging" is ontological suicide.[22] The reality we have encountered evidences that such an analogy is no longer part of the rhetorical convention and/or a customary portrayal of the algorithmic state of life only in critiques of the system; rather, it is an acute assessment of the current condition of being.

McDonaldization, a practice and culture that Ritzer explains as the all-inclusive snapshot of consumption and the true manifestation of rationality in the market system, is extending its territory virtually and transforming the platforms on which actors perform in the market.[23] Calling it a McDonaldization was an indirect and yet distinctive tribute to Max Webber's notion of rationalization and the subsequent pathological symptom at society level, control. As seen in everything else, the practice of McDonaldizing evolves into one that targets external stakeholders as part of the scheme. The original dimensions of McDonaldization included efficiency, calculability, predictability, and control. They were primarily intra- and interorganizational objectives and directives. The interface between organizations and consumers was continuously monitored and controlled, but the control never reached our cortex through screens and personal aides as it does now.

McDonaldization is much more than a mere metonymic expression of what was critiqued prior to the arrival of universal connectivity and more pertinent to the current state of consumerism that can be characterized as algorithmic consumer culture. It refers to

> a culture in which binaries (i.e., the visible and invisible, human and non-human, the ephemeral and permanent, the private and public, the simple and complex, participation and nonchalance, acquiescence and resistance, just to list some) in the conventional consumer-market dynamics are conjugated,

> imbricated, hybridized, and ultimately integrated into a system that perpetuates the cycle in favor of logic and control, which in turn, increasingly obfuscates the actual workings and makings of the system and often restricts the performativity of human actors and consumer agency in any network or assemblage.[24]

Staying normal has become the most elusive objective for consumers to achieve because either nothing seems normal or no one knows what normal has to connote. It has been a constant struggle for all not to be stigmatized for being less normal or "abnormal." At this algorithmic turn, however, normal is defined by what used to be subject to control by humans: the machine (i.e., algorithms and AI). For that reason, all of us as consumers are continuously in training to avoid losing the status of "modern consumer." Consumers who regularly interact with the machine maintain the status; the others have to justify their status subject to normalization.[25] Not having an *Amazon Prime* account, not having stayed at an *Airbnb* rental, and having no *Instagram* experience provide enough evidence for the machine to categorize the consumer as a dissident or a failure in the training regime. The machine nonetheless is restless to recruit and train more consumers through various touchpoints: reviews, online word of mouth, ads, in store point of sale, and social media. The only dominant ideology in the algorithmic consumer culture is conformity, and the machine's learning curve is rather gradual, if not horizontal.

While *normation* promotes a homogenization as a norm, *normalization* is a more sociotechnical practice that endorses malleability of the norm. An algorithmic normalization that shares properties of consumerization is a machine-led but essentially self-administered and self-inflicted process to stay normal.[26] Consumers, as a result, perceive the normalization-ridden culture of consumption as normal. Joining the culture no longer signifies "sellout" but indicates the inception of a "reversibility" pandemic. Dialectics is long gone, at least in the algorithmic consumer culture wherein the subject–object relationship is replaced with agency of agency. We, consumers, feel more empowered through algorithmic consumption practices because we are inundated with an unparalleled amount of information that is masked as the everlasting source of autonomy and sovereignty.

Consumer cyborgs are born every day because we never cease to incorporate technologies, media, and commercialized information with our minds. The posthuman epistemology has become more intelligible in theory as consumers' "techno-mind" blinds a clear view of the system to which we continuously respond.[27] The new epistemology is not about the relentless hybridization of subjects by algorithms but about consumers' conscious and voluntary subscription to the system with a belief that there will be a sufficient supply of distinction for subjectivity and identity.

What the algorithmic system promulgates is correspondingly the scientific method and process by which mid-range homogenization is recognized as the new norm in the market. In practice, consumers have inadvertently become a "reality machine" (or at least part of it) that (re)writes and (re)shapes the reality they always face with varying degrees of reflexivity. Sentiment analyses using social media postings unearth correlations and remote possibilities that would never be incorporated into business strategies otherwise. A simple Google search brings about a butterfly effect that can last indefinitely on the person or on others in the form of "recommendations." This "data-mined" self-identity is very graphic, linear, easily traceable, and all too docile. We, cyborgs, sometimes challenge and try to undermine the system, but it is often a futile battle that lasts only so long because there is no identifiable opponent. The more critical fight for consumers to win first is to learn how and when to unplug themselves.

IN STUDYING AGENTIAL

The dialectical model of consumption entails consumers' constant struggle to balance between the symbolic and the idiosyncratic. Consumers today, however, experience much less struggle because the boundary between the two has become blurred. This is a post-dialectical epoch wherein dialectical materialism is replaced with algorithmic materialism that eternalizes both the machine metaphor and the black box metaphor. Neoliberal governmentality as the currency for algorithmic culture does not invoke domineering measures to control; rather, it prefers "pasturage," with consumers always coming back to where they belong: the algorithmic system. Consumers in the market are programmed to think that they are always given opportunities to pursue autonomy, which is the very strategy algorithms are designed to implement at all times.

Consumers treat each moment of their lives as reality when they perform a role that algorithms allow them to play. Consumers also have their own modes of reality management, just as the market has different modalities of truth production (viewed as reality by consumers). This "veridiction" is more than the sum of consumers, algorithms, ideologies, and cultural orientations that overarch individual consumers' and marketers' reality production. To understand it, dialectical thinking must be renewed in terms of its scope and quality, because the consumer-algorithm dynamics often operate outside of the observable trajectory frequently discussed in consumer culture literature. One way the obscure dynamics can be analyzed is to adopt a genealogical method, especially for polyvalence of the power relations in algorithmic consumer culture.

Genealogy, as an attitude and the subsequent critique rather than a methodological template or manual, can offer keen insights for the ongoing cultural change in the market.[28] Traditional observation and intervention in reflexive science provide a "to do list" for scholars but do not necessarily provide guidance on how to approach an intricate phenomenon in which multiple objects and subjects are interacting without a common goal. Dialectical methods tend to blind one to novel and less demonstrable possibilities of truth-reality-making because binary oppositions and conflict resolutions always overshadow the underlying processes by which the subject-object relationship can be reversed. To understand less known ideologies and local identities that, in fact, sustain the market system, a method that actually embraces multiplicity is needed for future scholarly endeavors; and genealogy affords such an advantage to the upcoming analyses and discourses on algorithmic culture. What makes such an attitude a proper methodology is that, with it, one can deal with the complex, unstable, unbalanced, heterogenous, mobile, fragile, and (dis)continuous (re)distributions and (re)configurations of interactions and relationships frequently observed in the market.

FOR STAYING AGENTIAL

Given the epistemic disassembling and reassembling in recent times, the current state of algorithmic consumer culture, as a constellation of individual practices as well as the isomorphic end state of any successful enterprise, calls for a meaningful diagnosis to theorize the ever-growing and ubiquitous ambiguity in the algorithmic marketplace. The extant literature and public discourses help identify at least three stances academics, practitioners, and consumers can take responding to the highly obscure but ostensibly effective mechanism that governs almost everything without authority.

First, algorithms are taken for granted and even willfully trusted when the stance is *celebratory*. Subscribing to social media, shopping online, day trading, using dating apps, and googling require consumers to play certain roles while completing a variety of tasks. This stance also normalizes a voluntary denial of concerns and criticisms toward the black box that promotes the new normal for all.

Second, a *balance* between appropriating and being appropriated is also sought in the market. It is a strategic association with the new tech-intensive lifestyle. This approach espouses reflexive negotiation and subsequent attunement of our lifestyles. Turning off location services, blocking cookies, mindful use of social media, and overall cutback of exposure to the system all constitute strategic involvement in the inexorable cultural system that claims to offer convenience, entertainment, efficiency, precision, and, more importantly, the future.[29]

Third, a *critical* stance visualizes some conceivable issues witnessed and directly experienced by different actors in the market system, such as determination of creditworthiness, optimization of insurance premiums, biases from voice/facial recognition, and evaluation of individual performances at work, as well as the encroachment of the algorithm on our judicial system.[30] Culture in this day and age has been reimagined and reconfigured based on the spur of the moment trust in algorithms upon which current generations operate as free agents for the system. Trust is the only particular virtue such a cultural system promotes to maintain the system. Therefore, consumer culture should also be rearticulated when casual trust becomes the new prerequisite for all to maintain status as consumers in the market system. Subjectivity may have become an archaic value, replaced with a highly statistical proxy: an algorithmic approximation of a subject.

Finally, I argue that an ontological overhaul is inevitable for consumer subjects who conduct "businesses" with algorithms, and the process will be onerous because of the invasive nature of the new culture. Opportunism may be widespread in forthcoming discourses and theories, focusing on the conspiratorial facade of algorithms. Being opportunistic can be vastly advantageous or greatly precarious. However, the fear and risk involved in the reontologization of consumer subjects will be worth taking inasmuch as we continue to problematize fundamental characteristics of algorithms—namely opacity, messiness, inscrutability, hyper-legitimacy, hyper-functionality, and equal distribution of accountability—which are abused and overused as the source of authority of the "hollow" system.

NOTES

1. Deighton, John. "Big Data." *Consumption, Markets and Culture*, 22 (1) (2019), 68–73.
2. Jameson, Frederick. *Archaeologies of the Future: The Desire Called Utopia an Other Science Fiction*. New York: Verso (2007).
3. Giddens, Anthony. *The Consequences of Modernity*. Cambridge: Polity Press (1990); *Modernity and Self-Identity*. Cambridge: Polity Press (1991); *The Transformation of Intimacy*, Cambridge: Polity Press (1992).
4. Giddens, Anthony. *The Constitution of Society: Outline of the Theory of structuration*. Cambridge: Polity Press (1984).
5. Mauss, Marcel. *Sociology and Psychology*. London: Routledge (1979).
6. Schilbrack, Kevin. *Thinking through Rituals: Philosophical Perspectives*. New York: Routledge (2004).

7. Foucault, Michel. *The Birth of Biopolitics*. New York: Palgrave Macmillan (2008).

8. Foucault, Michel. *Society Must Be Defended*. New York: Picador (2003).

9. Foucault, Michel. *Security, Territory, Population*. New York: Palgrave Macmillan (2007).

10. Foucault, *Security, Territory, Population*.

11. Thomas, Suzanne L., Dawn Nafus, and Jamie Sherman. "Algorithms as Fetish: Faith and Possibility in Algorithmic Work." *Big Data and Society* (January 2018) doi:10.1177/2053951717751552.

12. Bardhi, Fleura and Giana M. Eckhardt. "Liquid Consumption." *Journal of Consumer Research*, 44 (3) (2017), 582–597; DuFault, Beth L. and John W. Schouten. "Self-Quantification and the Dataprenuerial Consumer Identity." *Consumption, Markets and Culture* (2020): 290–316.

13. Gopaldas, Ahir. "Marketplace Sentiments." *Journal of Consumer Research*, 41 (4), 995–1014 (2014).

14. Wilson-Barnao, Caroline. "How Algorithmic Cultural Recommendation Influence the Marketing of Cultural Collections." *Consumption, Markets and Culture*, 20 (6), 559–574 (2017).

15. Deighton. "Big Data," 68–73.

16. Leach, Edmund. *Culture and Communication*. Cambridge: Cambridge University Press (1976).

17. Gergen, Kenneth J. *The Saturated Self*. New York: Basic Books (1991).

18. Kopytoff, Igor. "The Cultural Biography of Things: Commoditization as Process." in *The Social Life of Things*, edited by Arjun Appadurai. Cambridge: Cambridge University Press (1986), 64–94.

19. Barthes, Roland. *Mythologies*, trans. Annette Lavers. London: Cape (1972).

20. Saussure, Ferdinand de. *Course in General Linguistics*, trans. Roy Harris. Peru, IL: Open Court (2000).

21. Kozinets, Robert V. "Can Consumers Escape the Market? Emancipatory Illuminations from Burning Man." *Journal of Consumer Research*, 29, 20–38 (2002).

22. Giddens, *The Consequences of Modernity*.

23. Ritzer, George. *The McDonaldization of Society*. Thousand Oaks, CA: Pine Forge Press (1993).

24. Hong, SoonKwan. "Algorithmic Consumer Culture." in *Research in Consumer Culture Theory Conference,* edited by G. Patsiaouras, J. Fitchett and A.J. Earley, Vol. 3. Leicester (2020).

25. Foucault, Michel. *Discipline and Punish: The Birth of the Prison*, trans. Alan Sheridan. New York: Vintage (1977) and *Power/Knowledge: Selected Interviews and Other Writings 1972–77*, edited by Colin Gordon. New York: Pantheon (1980).

26. Foucault, *Discipline and Punish*.

27. Giesler, Markus and Alladi Venkatesh. "Reframing the Embodied Consumer as Cyborg: A Posthumanist Epistemology of Consumption." *Advances in Consumer Research*, 32, 661–669 (2005).

28. Foucault, Michel. *The History of Sexuality Volume 1: An Introduction*. New York: Random House, Inc. (1978).

29. Vaidhyanathan, Siva. *Antisocial Media: How Facebook Disconnects Us and Undermines Democracy*. New York: Oxford University Press (2018).
30. O'Neil, Cathy. *Weapons of Math Destruction: How Big Data Increases Inequality and Threatens Democracy*. New York: Penguin Random House (2017).

BIBLIOGRAPHY

Bardhi, Fleura and Giana M. Eckhardt. 2017. "Liquid Consumption." *Journal of Consumer Research*, 44 (3), 582–597.
Barthes, Roland. 1972. *Mythologies*, trans. Annette Lavers. London: Cape.
Deighton, John. 2019. "Big Data." *Consumption, Markets and Culture*, 22 (1), 68–73.
DuFault, Beth L. and John W. Schouten. 2020. "Self-Quantification and the Dataprenuerial Consumer Identity." *Consumption, Markets and Culture*. 290–316 doi:10.1080/10253866.2018.1519489
Foucault, Michel. 1977. *Discipline and Punish: The Birth of the Prison*, trans. Alan Sheridan. New York: Vintage.
———. 1978. *The History of Sexuality Volume 1: An Introduction*. New York: Random House, Inc.
———. 1980. *Power/Knowledge: Selected Interviews and Other Writings 1972–77*, edited by Colin Gordon. New York: Pantheon.
———. 2003. *Society Must Be Defended*. New York: Picador.
———. 2007. *Security, Territory, Population*. New York: Palgrave Macmillan.
———. 2008. *The Birth of Biopolitics*. New York: Palgrave Macmillan.
Galloway, Alexander R. 2012. *The Interface Effect*. Cambridge: Polity Press.
Gergen, Kenneth J. 1991. *The Saturated Self*. New York: Basic Books.
Giddens, Anthony. 1984. *The Constitution of Society: Outline of the Theory of structuration*. Cambridge: Polity Press.
———. 1990. *The Consequences of Modernity*. Cambridge: Polity Press.
———. 1991. *Modernity and Self-Identity*. Cambridge: Polity Press.
———. 1992. *The Transformation of Intimacy*. Cambridge: Polity Press.
Giesler, Markus. 2008. "Conflict and Compromise: Drama in Marketplace Evolution." *Journal of Consumer Research*, 34, 739–753.
Giesler, Markus and Alladi Venkatesh. 2005. "Reframing the Embodied Consumer as Cyborg: A Posthumanist Epistemology of Consumption." *Advances in Consumer Research*, 32, 661–669.
Gopaldas, Ahir. 2014. "Marketplace Sentiments." *Journal of Consumer Research*, 41 (4), 995–1014.
Hong, SoonKwan. 2020. "Algorithmic Consumer Culture." in *Research in Consumer Culture Theory Conference*, edited by G. Patsiaouras, J. Fitchett and A.J. Earley, Vol. 3. Leicester.
Jameson, Frederick. 2007. *Archaeologies of the Future: The Desire Called Utopia and Other Science Fiction*. New York: Verso.
Kopytoff, Igor. 1986. "The Cultural Biography of Things: Commoditization as Process." in *The Social Life of Things*, edited by Arjun Appadurai, pp. 64–94. Cambridge: Cambridge University Press.

Kozinets, Robert V. 2002. "Can Consumers Escape the Market? Emancipatory Illuminations from Burning Man." *Journal of Consumer Research*, 29, 20–38.

Lash, Scott. 2007. "Power after Hegemony: Cultural Studies in Mutation?" *Theory, Culture and Society*, 24 (3), 55–78.

Leach, Edmund. 1976. *Culture and Communication*. Cambridge: Cambridge University Press.

Mauss, Marcel. 1979. *Sociology and Psychology*. London: Routledge.

O'Neil, Cathy. 2017. *Weapons of Math Destruction: How Big Data Increases Inequality and Threatens Democracy*. New York: Penguin Random House.

Ritzer, George. 1993. *The McDonaldization of Society*. Thousand Oaks, CA: Pine Forge Press.

Saussure, Ferdinand de. 2000. *Course in General Linguistics*, trans. Roy Harris. Peru, IL: Open Court.

Schilbrack, Kevin. 2004. *Thinking through Rituals: Philosophical Perspectives*. New York: Routledge.

Thomas, Suzanne L., Dawn Nafus, and Jamie Sherman. 2018. "Algorithms as Fetish: Faith and Possibility in Algorithmic Work." *Big Data and Society*. doi: 10.1177/2053951717751552

Vaidhyanathan, Siva. 2018. *Antisocial Media: How Facebook Disconnects Us and Undermines Democracy*. New York: Oxford University Press.

Wilson-Barnao, Caroline. 2017. "How Algorithmic Cultural Recommendation Influence the Marketing of Cultural Collections." *Consumption, Markets and Culture*, 20 (6), 559–574.

Chapter 3

Monoculturalism, Aculturalism, and Post-Culturalism

The Exclusionary Culture of Algorithmic Development

Ushnish Sengupta

The branch of the tech industry engaged in algorithm development typically—by which I mean here the design and development of algorithms in the context of software applications—views itself as "acultural," that is, as independent of culture or cultural bias and as transcending cultural differences in developing the future. This chapter argues that algorithm development is instead deeply rooted in narrow cultural practices that exclude genuine input from a broad variety of participants and perspectives. Consequently, the negative impacts of algorithm development that result from algorithmic bias are experienced primarily by individuals and communities who are not meaningfully involved in the development of the algorithms that affect them. A deliberate and significant effort toward increasing cultural diversity in the design and development of algorithms is required to enable genuine input from a broader group of participants with more expansive ideas about development, which will enable addressing the concerns of more diverse communities and contribute a greater variety of innovative and appropriate solutions.

The chapter focuses on two separate, but intersectionally identifiable, cultural biases in algorithm development. The first cultural bias is related to the male-dominated, patriarchal culture of algorithm development. Under patriarchy, data from technology firms in the United States and Canada have repeatedly reproduced both the underrepresentation of women and an unwelcome culture of algorithm development for women. The establishment of gender responsive policies that would enable individuals across the gender spectrum to bring their whole selves to work in the industry without discrimination in hiring, work assignment, and promotion would improve the climate for women in algorithm development. There is a substantial body of research validating the economic and social benefits of closing the gender gap in industries and subsectors where there is an imbalance of gender representation.[1] By changing this patriarchal culture, the industry would benefit from a broader diversity of ideas and related innovation.

The second cultural bias is the dominance of an American and European culture of algorithm development. Influenced by history, geography, and economics, including colonialism and nationalism, the capacity to develop algorithms and to create the large technology companies that increasingly develop them is associated with a deeply embedded nationalistic view of citizens, governments, and technological progress. Current algorithm development training and employment patterns need to be analyzed critically as contributing to acculturation into these existing cultural orientations. Merely diversifying employee pools but maintaining the same cultural orientation inculcated in training will not solve deeply rooted cultural biases that contribute to limited labor market outcomes and disproportionate harm to particular ethnoracial groups. In other words, achieving a diversity of bodies within the current structures of the tech industry does not necessarily achieve diversity of thought, and it is diversity of thought that is required to broaden the cultural orientation of algorithm design and development.

PATRIARCHAL CULTURE AS MONOCULTURE IN ALGORITHM DEVELOPMENT

Understanding the cultural influences in algorithm development starts with understanding the patriarchal culture of the industry. The deep cultural foundations of algorithm development are tied to the social characteristics of the individuals and groups who have developed and led the industry. Without understanding the cultural situatedness of the industry, one cannot understand the blind spots that limit the industry's effective growth. First and foremost, that culture is monocultural. Evidence of its existence comes from broad surveys such as the *Global Gender Gap Report* completed by the World Economic Forum,[2] which documents a significant gender gap in the share of professionals with artificial intelligence (AI) skills: from a high of 82 percent in favor of men in countries like Germany and Mexico to a low—but still substantial—61 percent in countries such as Italy and Singapore. Two observations merit attention. First, this gender gap exists across countries and continents. It is not, therefore, a country-specific cultural issue, but a global, patriarchal, technology industry-wide issue. Second, even though the *Global Gender Gap Report* indicates that there is an increase in the rate of training in AI skills by both men and women across the world and that they are increasing at the same rate, there is no narrowing of the existing gender gap in employment, which does not portend well for correcting this imbalance in AI.

While gender disparity in the development of AI, one of the fastest-growing areas of algorithm development, can at least at a superficial level be perceived as a critical issue, there are multiple indicators that the narrow patriarchal culture of AI development can lead to negative impacts for both individuals and organizations. One example is the application of AI to the process of recruiting by *Amazon*.[3] As one of the largest companies in the world, *Amazon* receives a high volume of applications and resumes for available positions. As a company with technology development as one of its core competencies and as a leader in the application of AI, *Amazon* applied AI to its recruiting process. The application of AI would ideally make the process more efficient, enabling analysis of thousands of resumes at a lower cost and with greater effectiveness than human recruitment processes could accomplish. The use of algorithms to look for keywords in electronically submitted resumes is not particularly new. What was new in this case was the application of training data based on profiles of successful *Amazon* employees. Ideally the algorithm would analyze the resumes and predict "good fits" based on matching applicant patterns with individuals who were already successful employees. The AI-based recruiting system worked exactly as designed; it recommended candidates whose characteristics fit with individuals who were already successful at *Amazon*. The problem, of course, was that the AI-based recruiting system recommended candidates who were male, since individuals who were thriving at *Amazon* were predominantly male. The AI-based recruiting learned to recognize that one of the most common characteristics of individuals who have already been successful at *Amazon* was that they self-identified as men. More insidiously, it learned to identify and penalize resumes of female candidates from keywords such as "women's chess club captain" and to identify and penalize candidates who graduated from all-women's colleges.[4] In this way, the gender imbalance at *Amazon* was not only identified but amplified. Early on in the deployment of the AI-based recruiting system, employees of *Amazon* recognized this fault in the system and canceled the project. It is, however, worth stating the obvious: recruiting candidates that were only male, as recommended by the AI-based recruiting system, would *not* provide the company with the best talent, and over the longer run it would limit innovation and business success.

Patriarchal reproduction and enhancement in algorithm development are not specific to *Amazon*, but exist across the industry. In another example, Joy Boulamwini examined the ability of commercial facial recognition software applications, including a product sold by *Amazon*, to recognize the faces of a variety of people.[5] Boulamwini found that the facial recognition software was more accurate recognizing male faces than female faces, and most accurate for lighter male faces and least accurate for darker women's faces. Facial recognition software experts often blame existing datasets of faces as being the main reason for this bias in facial recognition between male and female and between lighter and darker faces. In other words, the training datasets typically contain a larger quantity of lighter male faces compared to other faces and, therefore, the facial recognition algorithm is better trained to recognize lighter male faces. However, it is insufficient to blame the dataset or even attempt to fix this cultural failure with a more inclusive dataset; the technological fix is a superficial and inadequate solution. A more appropriate and forward-looking cultural response would be to ask why the dataset selected overrepresented or underrepresented particular populations to begin with. A more culturally progressive path toward a solution would ask about the composition of the team developing the facial recognition software application, their beliefs and their values, and processes which cumulatively led to the biased data set and ultimate result. Fixing the data set or fixing the team to ensure more diversity will not necessarily solve the deeper cultural issues at the root of the problem. There is no substantial evidence indicating that an equal number of men and women or an equal number of people of color will necessarily produce different results.

There are many benefits to a diversity of people in any organization, but diversity of people does not necessarily equate to diversity of thought and values. For example, if all the software developers on a team come from similar socioeconomic backgrounds, went to similar educational institutions, and were therefore acculturated into a particular value system, the diversity of thought based on different values will be limited. Therefore, while it may be necessary to have diversity of people on an algorithm development team, it is not sufficient. It is also necessary to have the diversity of thought that comes from a diversity of values, and diversity of values comes from a diversity of cultures. Returning to the *Global Gender Gap Report* and the gender gap in the AI talent pool, one of the enlightening parts of the report is the difference in the AI talent gender gap across different industries.[6] The report indicates that although the AI talent gender gap is high in the Software and IT Services industry, it is lower in nonprofits, health care, and education industries. We know that nonprofit sector and public sector organizations are culturally distinct with different sets of values than private sector organizations. Further, the findings support the contention that cultural differences in these industries and organizations result not only in more equitable employment of women but also in a lower gender gap in those industries. While this is not an argument for turning private sector institutions into nonprofits or public sector organizations to address gender gaps in talent, it is an argument for private sector organizations to learn from the cultural practices of nonprofit and public sector organizations. The recommendation here is to identify the cultural practices of nonprofit and public sector organizations that often make them more likely to be places where more women want to apply their skills and experience, including women who are on the leading edge of the algorithm development industry with AI skills.

CORPORATE CULTURE AND ACULTURAL ALGORITHM DEVELOPMENT

In this section, I discuss aspects of the branch of the tech industry engaged in algorithm development which typically views itself as "acultural." The term "acultural" is utilized here as described by Gershon and Taylor:

> The people actively carrying out the work of the institution take the institutional context for granted and do not consider its particular forms, forums, patterns, and practices to be cultural. Instead, in these institutional contexts, a select few outsiders get defined as the bearers of culture.[7]

Many tech companies and institutions engaged in algorithm development, similarly do not consider organizational patterns and practices to be cultural. Corporate culture in technology companies needs to be viewed as part of the acculturation process into the dominant culture of algorithm development. Trying to address cultural issues in algorithm development by diversifying the employee pool, but propagating and maintaining the same dominant cultural basis, will not solve deeply rooted culturally dependent issues. As I have argued, building teams with diverse groups of individuals without examining the cultural biases of algorithm development will continue to produce inequitable results related to the industry, and whether and how one is excluded or excluded by corporate culture is one of the ways this comes about.

In countries with racialized minority populations such as the United States and Canada, patriarchal and Eurocentric-based racial imbalance is particularly evident in the underrepresention of racial minorities in algorithm development. The impacts are intersectional, in that they do affect not only gender and race but also individuals differently at the intersection of these facets of identity. The Brookfield Institute for Innovation + Entrepreneurship in Canada provides evidence for these intersectional differences. In "Who Are Canada's Tech Workers?" they report that particular racialized minorities in Canada are disproportionately underrepresented.[8] Black, Filipino, and Indigenous communities are particularly underrepresented in technology occupations in Canada. Moreover, within each racialized community, there are wide gaps in the participation and pay for male and female technology workers. For instance, Chinese, West Asian, and South Asian *men* have a higher participation rate than the reference "not a visible minority," but *none* of these racialized communities achieve greater average pay than "not a visible minority." Further, women in every racialized category, including "not a visible minority," have both a lower participation rate and a lower average pay in technology occupations compared to men in the same group. In other words, Chinese women have both a lower participation rate and a lower average pay than Chinese men in technology occupations; South Asian women have both a lower participation rate and a lower average pay than South Asian men in technology occupations; Black women have both a lower participation rate and a lower average pay than Black men in technology occupations, etc.

Attributing participation rate to the gap in average pay, or assuming that more women would participate if they were paid equally for equal work, are overly simplistic assumptions to make about how companies and organizations might counter then institutionalized acculturism. While it is true that lower average pay may discourage many racialized women from participating at the same level as men in their community, equalizing pay between men and women will not alone solve the deeper cultural issues. We must ask why average pay for women within each racial group and overall is lower, and what cultural issues—in addition to equalizing pay—can be addressed in making algorithm development and the technology industry more broadly a welcoming place to recruit, retain, promote, and be responsive to racialized female talent.

"The Illusion of Asian Success: Scant Progress for Minorities in Cracking the Glass Ceiling from 2007–2015," a multiyear study produced by the think tank Ascend, provides additional insight into the complexity of intersectional representation in algorithm development.[9] Analyzing technology companies in the United States, the study reports on the relative progress and advancement of racialized minority men and women in the industry. The report finds that although White women have made progress gaining executive positions in proportion to their representation in the entry-level workforce across companies, Asian and Hispanic women have made little progress, and Black women have made no progress at all. Additionally, Asian women and Black women have the lowest ratio of executive to entry-level staff compared to all other groups. In other words, Asian, Hispanic, and particularly Black women face a "Glass Ceiling" in being promoted to executive positions in these companies. Therefore, the issue is not a simply one of equal pay. In addition to equal pay, the cultural factors that limit the progress of Black, Hispanic, and Asian women within technology companies need to be examined. The data also show that the ratios of executive to professional staff for Black, Hispanic, and Asian men are higher when compared to similar groups of women, but are still lower than White men and women. These data support the existence of a glass ceiling preventing racialized minorities from reaching executive positions in technology companies, suggesting the persistence of a cultural assumption that Black, Hispanic, and Asian employees do not have the skills to be executives. Even though groups such as Asians are becoming a greater proportion of technology company employees in the United States, they are not advancing to executive positions at the same rate as other groups. This result may be based on the inappropriately biased view that Asians are a "model minority," whereby the community is labeled as being good at specific subjects—in this case as good at mathematics, and great as employees of technology firms—but bad at others—in this case as not having the skills to become executives. In a different but equally inappropriate and biased view, Black individuals are often labeled as not being good at mathematics and not great as employees, particularly in technology firms. In "The Illusion of Asian Success," there is a corresponding steady and long-term decrease in the proportion of Black individuals hired and employed as professionals in technology companies.[10] These cultural biases and intersectional stereotypes are typically hidden, disguised by the "apparent" commitment to aculturalism in algorithm development.

The algorithm development industry as a whole promotes the idea of a meritocracy, a cultural concept firmly entrenched in technology industry folklore. In its ideal form, the idea is a noble one: individuals are hired, promoted, and rewarded according to the merits of their work rather than who they are in terms of nationality gender, race, sexual orientation, disability, and so on. This narrative of meritocracy serves the algorithm development industry well, as it needs and attracts talented individuals from across the world to work on its projects. Individuals who aspire to work for, be promoted within, and make it to the top of algorithm development organizations need to believe in the idea of meritocracy. These individuals need to believe that it is only the merits of their work that matter and that the country they come from and their cultural background do not factor into hiring, promotion, and compensation decisions. By promoting the belief that the algorithm development industry is a leader in and an island of meritocracy, organizations are able to attract the brightest and best talent from across the world. Although stories of successful individuals from diverse cultural backgrounds are often highlighted, such as the successful immigrant who has either started successful companies or has become an executive, the data tell a different story.

This chapter has highlighted different surveys from two different, but culturally similar countries—Canada and the United States—where the patterns of employment and promotion clearly reflect the cultural influences and societal stereotypes that exist in these two countries. Although there are individuals from different cultural backgrounds who have been successful in algorithm development organizations, the broader pattern across the population is one of dominant cultural views influencing hiring, promotion, and compensation decisions. Successful exceptions do not define the overall pattern; they are simply exceptions. Therefore, although the algorithm development industry promotes a narrative of meritocracy and aculturalism, it remains firmly situated in the national cultures with their biases and stereotypes.

A final example demonstrates the degree to which algorithm development organizations are steeped in the cultural context of the countries they are based in, rather than being acultural entities. In this example, I consider some of the cultural beliefs of three founders of algorithm development organizations. These founders of algorithmic development organizations are products of the dominant culture and related subcultures they are enmeshed in. They provide the initial strategic direction of these organizations and set their cultural character and tone. The first founder is Richard Stallman, who established a number of algorithm development organizations. He resigned from his positions at MIT and the Free Software Foundation after expressing controversial views on sexual assault of women via email, views that were linked to a longer-term pattern of patriarchal behavior.[11] This example is particularly disturbing as Stallman started and was involved in a number of educational organizations and was a well-known leader in the industry. The Stallman example reinforces the point being made that cultural issues pervade algorithm development institutions, including for-profit, nonprofit, and public sector organizations. The second founder is William Shockley one of the recognized founding fathers of Silicon Valley, who promoted a racist eugenics point of view.[12] Shockley not only set the strategic direction and cultural tone of his own company but was influential in setting the strategic direction and cultural tone for an entire industry in California's Silicon Valley. The cultural trajectory of Silicon Valley could have been different, if the initial founder had different and more progressive cultural views. The third example is Damien Patton, the founder of up and coming algorithm development organization Banjo. He resigned as CEO when his past ties to the KKK and involvement in a drive-by shooting of a Synagogue were exposed. The example of Patten is important for two reasons. First it shows that the initial race-based cultural influences of Silicon Valley founders such as Shockley were not simply a reflection of the views of a few people in the 1950s and 1960s, but were part of a chain of cultural and subcultural influences that have continued into the 1990s and the 2000s. This is not to say that the majority of Silicon Valley entrepreneurs are racist, but that key figures hold racist cultural views that contribute to shaping and damaging the industry and society. Second, the Patten case is important in that Banjo developed AI-based surveillance technology that was sold to several governments. Many in the AI community will describe the technology as "neutral," as acultural. However, as this chapter has argued, algorithms are developed by people, and biases in algorithms are consciously or unconsciously shaped by the people developing the algorithms. When the founder of an AI-based surveillance technology company holds explicit racist beliefs and values, it calls into question the purported neutrality of algorithm development within that organization and suggests the presence of biases in the development and implementation of their algorithms.

The analysis of the cultural basis of algorithm development necessarily extends beyond the surface layer narratives of meritocracy to a more critical examination of the cultural values embedded in algorithm development. In turn, the cultural lens provides a multilayered examination including processes, people, ownership of organizations, and decisions that promote dominant cultural stereotypes and biases at the expense of others. As I have argued, the organizational culture that currently drives the design and implementation of algorithms is deeply biased and narrowly constituted. Next, the chapter draws on the deep multilayered cultural analysis of the field of algorithm development education, surprisingly a subfield that is not often analyzed for cultural underpinnings even though the field of education more broadly is frequently analyzed for cultural bias.[13]

POST-CULTURALISM, EDUCATION, AND ALGORITHM DEVELOPMENT

Industry practices drive educational practices, and learning algorithm development takes place primarily in educational institutions. This section argues that educational institutions that generate and perpetuate the growing body of knowledge in algorithm development are firmly culturally situated, in opposition to the belief held by many of these educational institutions that what they teach is "post-cultural." I use the term post-cultural here as described by Witte in order to signal organizations whose cultures are post-national cultures.[14] In other words, I argue that educational institutions tend to erroneously subscribe to the belief that their approaches toward teaching algorithmic development are not inflected by the national cultures. I focus on the algorithmic game development industry as an example of the post-cultural belief in algorithm development. Algorithmic game development provides a rich context to understand the cultural basis as there is a tight coupling between the industry and the educational institutions training employees for the industry. As described by Weststar, the video game development community must be viewed as an occupational community,[15] a community that has a number of common characteristics, and a community that has deep connections with both education and industry. Algorithmic game development does not involve developing a software algorithm as a tool in and of itself but rather subordinates the algorithm in service of narratives, plot lines, characters, reward systems, and more recently celebrity gamers. Therefore, a cultural lens can be used to examine what the common narratives and plotlines are, who are the characters and are they subject to stereotypes, and how do the reward systems both for gamers and game makers amplify culture. The gaming industry, including its leading educational institutions, is an international industry with popular games sold or played by millions across the world. Educational institutions teaching game development consider themselves post-cultural as they include international students, and graduates who work in different countries. Yet the main protagonists, plotlines, and reward systems in algorithmic games have been found to be racist,[16] sexist,[17] homophobic,[18] and ableist.[19] For example, popular algorithmic games have portrayed people from Arab countries as terrorists,[20] which is a part of the long-standing cultural tradition of Orientalism in algorithmic game development.[21] The demonization of Arab and Muslim characters in games as the "other" is a Eurocentric cultural tradition that is continued in North America. These characteristics of popular algorithm-based games can be accounted for by the stereotypes of Eurocentrism that structure western nationalistic cultures. Learning algorithmic game development at educational institutions therefore includes not only technical programming skills but also sociocultural aspects of game design that are based on Eurocentric, male, heterosexual, and ableist cultural assumptions.

Competency in algorithm development can be attained by individuals through self-learning and experience, unlike other regulated professions such as engineering, medicine, or law. The ability to develop skills through self-learning and experience contributes to the myth of meritocracy in algorithm development. Most people, however, particularly at early career stages, need the scaffolding and more formalized learning processes provided by educational institutions to achieve a level of competency that is valued by employers. The educational system, which includes formal educational institutions and independent learning processes which depend on the knowledge generated by educational institutions, thus becomes a site for cultural reproduction. There is a canon for every field of study, a set of orthodox values that forms the core of the field, and that is typically resistant to unorthodox ideas that can change the values and beliefs in the field. For example, there is the mainstream field of economics that dominates over marginalized ones: heterodox economics that take feminist, ecological, decolonial, and other perspectives.[22] Similarly, in management, the field of Critical Management Studies provides formal critiques of many management paradigms.[23] At the present, in spite of development of critical literature in game studies, there is still insufficient attention to critiques of post-cultural assumptions of algorithmic development.

FIRST, DO NO HARM

Moving toward solutions for the issues of cultural biases in algorithm development necessitates first, and foremost, an outline of the potential harms of the current monocultural, acultural, and post-cultural views of algorithm development. When it comes to algorithm development, one of these core tenets of the field is technosolutionism.[24] One of the facets of technosolutionism is that the benefits of using new technological solutions to address the majority of problems are always seen as being greater than the costs or risks. This is not perhaps different from an aspiring lawyer internalizing the belief that learning, applying, and changing the law will solve a larger number of problems. Where technosolutionism differs from the study of law is that it assumes that technology solutions are value neutral, that the implementation of technological solutions based on their gender, income, race, and ethnicity. The Institute for Women's Policy Research has completed a study indicating that women will face disproportionately negative consequences from AI, particularly in the area of employment.[25] This chapter has already described how there is an existing, and ongoing gap between the proportion of men and women working in AI-related jobs. The Institute for Women's Policy Research report predicts that women will lose more jobs from the implementation of AI since the jobs that are more likely to be automated are more likely to be held by women. Therefore, the design, development, and implementation of algorithms is never value or culture neutral, it has different impacts on different groups, it is firmly rooted in a culture and it is not neutral as purported by many supporters of technosolutionism. Algorithm development is deeply embedded in a patriarchal and Eurocentric culture that informs the field and the industry.

Silicon Valley in the United States in particular has a reputation for providing the ideal context for development of disruptive firms. Disruptive firms create new products and services that transform entire industries, and create new technology-based subindustries. The companies are often different from existing companies in terms of organizational structure and management processes. The similarity in the companies is the small set of educational institutions which generate the majority of Silicon Valley entrepreneurs. Research by Pitchbook indicates that the top 10 schools generating the most entrepreneurs have not changed substantially over the last few years.[26] In other words, the majority of technology company founders learn the techniques from the same set of educational institutions. Although this phenomenon does not result in organizational isomorphism, it does result in cultural reproduction of the existing culture of algorithm development.

This brings us to the last major argument in this chapter that the tendencies of monoculturalism, aculturalism, and post-culturalism in algorithm development results in harm to particular underrepresented groups. It is important to address who and what benefits and who and what is harmed by the current cultural orientation of algorithm development. Virginia Eubanks has illustrated how algorithms developed for the purpose of delivering government services more efficiently to save the government money typically target and negatively impact low income, racialized families.[27] Current research is building a catalogue of these impacts. For example, Safiya Noble has documented how search engines reinforce inappropriate, racist stereotypes rather than mitigate or counter them.[28] Shoshana Zuboff has shed light on how the algorithm industry, including social media companies, collect enormous amounts of data for the purpose of generating revenue.[29] Russell Brandom has identified the implementation of algorithms by *Facebook* that selectively display real estate advertisements as reproducing the redlining policies of earlier decades that exclude Black individuals from buying property in certain geographical areas.[30]

In response to these patterns of benefit and harm, several interrelated strategies can be implemented. The first has to do with the question of representation in design and development teams and structures. If algorithm development teams, leadership, and organizational structures within which the work is taken on, designed, implemented, and rewarded were culturally diverse, they would likely have less harmful impacts on underrepresented groups. If, for example, an algorithm development team were evenly divided between men and women, the team would be less likely to build a system that was disproportionately harmful to women. Similarly, if an algorithm development team included racialized individuals from a particular group, the team would be less likely to develop an algorithm harmful to that particular group. Diverse team members would be able to identify the potential harm or risks before they occurred based on their lived experience of some of these harms

However, as this chapter has argued, the cultural diversity of individuals' backgrounds is a necessary but insufficient corrective. Achieving a diversity of bodies within current structures does not necessarily achieve diversity of thought and value, which is the second corrective that can be implemented. Ultimately, diversity of thought and value is required to challenge and transform the cultural biases of algorithm design and development. Thus, for example, if the organizational culture within which the team operates rewards the development of an algorithm that disproportionately harms women, it isn't likely to matter if the development team is diverse.

But how can algorithm design achieve diversity of thought? Rewards, particularly in private sector organizations, matter; and rewards for algorithm development teams and individual team members, are typically tied to evidence of increasing revenues for the organization.[31] Eubanks has argued that rewards in public sector organizations for algorithm development teams are typically tied to decreasing costs for the organization.[32] Therefore, a variety of different cultural perspectives are required on algorithm development teams. If team members all graduated from similar educational institutions and have similar work experience, it will be difficult to achieve diversity of thought. Even a racialized individual or an immigrant with education and work experience similar to the rest of the team is unlikely to offer a significantly different cultural perspective. Algorithm development across the world, as described earlier, has taken on a monocultural character, as the same field canon is followed by educational institutions around the world. They tend to replicate the organizational structures found in Silicon Valley, since they want to emulate the financial success of Silicon Valley firms.

As eloquently expressed by Audre Lorde, "The Master's Tools Will Never Dismantle the Master's House." What is required to diversify the algorithm development industry is identifying, employing, and rewarding individuals with genuinely different cultural backgrounds and ways of thinking. To do this requires looking to other fields of knowledge that have made contributions to the field of algorithm development, including psychology, sociology, critical gender studies, critical race studies, and so on.

FUTURES

At this point, a number of arguments have been made about the monocultural, acultural, and post-cultural assumptions in algorithm development. These assumptions have been challenged in this chapter by highlighting the hidden but essential cultural basis of algorithm development. The aspect of harm for particular communities from current practices of algorithm development has been explained in the previous section. The solution to the issues is not to remove the cultural basis for algorithm development, attempting to do so simply makes the cultural basis more invisible and amplifies the incorrect assumptions of monocultural, acultural, and post-cultural algorithm development. The solution is to apply plural cultures to algorithm development, specifically bringing different cultural perspectives that are different from the dominant Eurocentric culture, such as Afrofuturism and Indigenous futurism. The term "pluralism" is preferred here rather than multiculturalism, since multiculturalism has recently taken on a meaning of superficialness, of an appreciation of dress and dance and diet rather than substantial examination of cultural differences and lived realities. As eloquently expressed by Audre Lorde, "The Master's Tools Will Never Dismantle the Master's House." The cultural issues related to algorithm development will not be solved by currently entrenched monocultural, acultural, and post-cultural thinking in algorithm development. We need to look outside these cultures, to cultures different from Eurocentric culture such as Indigenous futurism[33] and Afrofuturism, for example, of how algorithm development can be implemented differently if there was a greater cultural plurality in the development of algorithms.

Indigenous futurism, as defined by Dillon,[34] describes a future where Indigenous people have an essential role in the future, where Indigenous communities have not been incorporated into dominant cultural futures, but contribute to the future from an Indigenous knowledge base. Indigenous futurism can be seen in algorithmic game development: for example, the perspective used in the development of games by Indigenous game developers on how Indigenous characters are portrayed is significantly different from non-Indigenous game developers[35] Indigenous futurism can be appropriately employed in game development by non-Indigenous game developers under the guidance and leadership of Indigenous communities. Madsen discusses algorithmic game development by a non-Indigenous organization, telling an Indigenous story, shaped by more than forty Indigenous elders during development of the game.[36] Educational institutions can also play a role in incorporating Indigenous futurism into algorithmic game development. Lameman and Lewis,[37] for example, discuss a partnership with a Canadian post-secondary institution where Indigenous educators guided the direction of the game development project that involved training Indigenous youth to develop games. Indigenous leadership in algorithmic development can therefore guide the development and direction of the project, using Indigenous values and philosophies, to achieve outcomes relevant for Indigenous and non-Indigenous communities. In another example, a technology platform, Second Life, has been used to create and develop virtual Indigenous spaces that utilizes existing platform capabilities and develops them further using Indigenous knowledge.[38] This set of examples of Indigenous futurism in different forms of algorithm development exemplify the principle that Indigenous futurism is not solely for the benefit of Indigenous people, nor is it exclusively about Indigenous people developing technology on their own. There are many models that involve leadership by Indigenous communities in reflecting their values in the direction of technology development.

Afrofuturism brings another cultural perspective to pluralistic algorithm development, from the pioneering work of Octavia Butler in defining Afrofuturism,[39] to the more recent cinematic representation of Afrofuturism expressed in the movie Black Panther by director Ryan Coogler.[40] Use of Afrofuturism in algorithm development can be found in education, where, for example, teachers have used Afrocentric knowledge to teach coding for mathematics applications,[41] or culturally responsive pedagogy to develop peer leaders among African American youth learning to code.[42] Other examples on a continental scale include African artificial intelligence, which utilizes AI for solving problems of interest on the African continent. On a local scale, in the United States a promising trend is the opening of Black incubators led by Black individuals that provide training on algorithm development coupled with entrepreneurship skills. For example, the Inclusive Innovation Incubator in Washington DC led by Black entrepreneur Aaron Saunders, which trains underrepresented youth in algorithm development including games, empowers them to become creators beyond being users or players, to embed their own stories into algorithm development with additional training to make their products commercially viable. The Inclusive Innovation Incubator is particularly promising as it is not an organization operating in a silo, but is connected to a city wide plan to increase involvement of underrepresented groups in algorithm development.

Algorithm development therefore can have different possible paths, and different benefits and harms based on not only who develops them but also how they are designed and implemented from different cultural bases. This chapter addresses a gap in the prevalent literature on multilayered cultural analysis of algorithm development, building on similar analysis in other fields. This chapter intends to contribute toward solutions for the culturally based, long-term inequities in outcomes in algorithm development for particular groups, and reduction of the disproportionate harms that are the result of the current algorithm development, design, and implementation practices. At the same time, this chapter presents as a solution pluralistic algorithm development education and practice, which incorporates Indigenous, Afrocentric, and other knowledges to create a future that is not monocultural, acultural, or post-cultural, but recognizes the valuable contributions of different visions of the future from different communities. Indigenous futurism and Afrofuturism are presented as central goals to be worked toward, rather than extending a dominant Eurocentric base of algorithm development toward the margin.

NOTES

1. Woetzel, J. Manyika, J., Dobbs, R.,Madgavkar, A., Ellingrud, K., Labaye, E., Devillard, S., Kutcher, E, Krishnan, M. 2015. "How advancing women's equality can add $12 trillion to global growth." McKinsey & Company.Retrieved from: https://www.mckinsey.com/featured-insights/employment-and-growth/how-advancing-womens-equality-can-add-12-trillion-to-global-growth.
2. World Economic Forum. *The Global Gender Gap Report.* (2018). Retrieved from https://www.weforum.org/reports/the-global-gender-gap-report-2018.
3. Jeffrey Dastin, Amazon scraps secret AI recruiting tool that showed bias against women. *Reuters* (October 9, 2018).

4. Ibid.
5. Joy Boulamwini. *Actionable Auditing: Coordinated Bias Disclosure Study.* MIT Media Lab (2019).
6. World Economic Forum, *The Global Gender Gap Report.*
7. Gershon, I. and J. S. Taylor. "Introduction to 'in focus: Culture in the spaces of no culture." *American Anthropologist,* 110 (2008), 417–421.
8. Vu, Lamb & Zafar, A. "Who are Canada's tech workers?" *Brookfield Institute for Innovation + Entrepreneurship* (2019).
9. Gee, B., and D. Peck. "The illusion of Asian success: Scant progress for minorities in cracking the glass ceiling from 2007–2015." *Ascend* (2019) Retrieved from https://cdn.ymaws.com/www.ascendleadership.org/resource/resmgr/research/theillusionofasiansuccess.pdf.
10. Ibid.
11. Lee, T. B. "Richard Stallman leaves MIT after controversial remarks on rape." *Arts Technica* (2019).
12. Rosenberg, S. "Silicon Valley's first founder was its worst." *Wired* (July 19, 2017) Retrieved from https://www.wired.com/story/silicon-valleys-first-founder-was-its-worst/.
13. Cupples, J., and R. Grosfoguel. (eds.). *Unsettling Eurocentrism in the Westernized University.* Routledge (2018).
14. Witte, A. "Making the case for a post-national cultural analysis of organizations." *Journal of Management Inquiry,* 21 (2000), 141–159. doi: 10.1177/1056492611415279.
15. Weststar, J. "Understanding video game developers as an occupational community." *Information, Communication and Society,* 18(10) (2015), 1238–1252.
16. Leonard, D. J. "Not a hater, just keepin'it real: The importance of race-and gender-based game studies." *Games and Culture,* 1(1) (2006), 83–88.
17. Vermeulen, L., M. V. Abeele, and S. Van Bauwel. "A gendered identity debate in digital game culture." *Press Start* 3(1) (2016), 1–16.
18. Condis, M. "No homosexuals in Star Wars? BioWare, 'gamer' identity, and the politics of privilege in a convergence culture." *Convergence,* 21(2) (2018), 198–212.
19. Crooks, H. R., and S. Magnet. "Contests for meaning: Ableist rhetoric in video games backlash culture." *Disability Studies Quarterly,* 38(4) (2018).
20. Saleem, M., and C. A. Anderson. "Arabs as terrorists: Effects of stereotypes within violent contexts on attitudes, perceptions, and affect." *Psychology of Violence,* 3(1) (2013), 84.
21. Šisler, V. "Digital Arabs: Representation in video games." *European Journal of Cultural Studies,* 11(2) (2008), 203–220.
22. Tae-Hee, J. "Frederic S. Lee and his fight for the future of heterodox economics." *PSL Quarterly Review,* 69(278) (2016), 267-278.
23. Prasad, A., and A. Mills. "Critical management studies and business ethics: A synthesis and three research trajectories for the coming decade." *Journal of Business Ethics* 94 (2010), 227–237 .

24. Gooch, D., M. Barker, L. Hudson, R. Kelly, G. Kortuem, J. V. D. Linden, and J. Mackinnon. "Amplifying quiet voices: Challenges and opportunities for participatory design at an urban scale. *ACM Transactions on Computer-Human Interaction (TOCHI),* 25(1) (2018), 1–34.

25. Hegewisch, A., C. Childers, and H. Hartmann. *Women, Automation, and the Future of Work.* Institute for Women's Policy Research (2019).

26. Pitchbook. PitchBook Universities (2019) Retrieved from https://pitchbook.com/news/articles/pitchbook-universities-2019.

27. Virginia Eubanks. *Automating Inequality: How High-Tech Tools Profile, Police, and Punish the Poor.* St. Martin's Press (2018).

28. Safiya Noble. *Algorithms of oppression: How Search Engines Reinforce Racism.* NYU Press (2018).

29. Shoshana Zuboff. *The age of Surveillance Capitalism: The fight for a Human Future at the New Frontier of power.* Profile Books (2019).

30. Russell Brandom. "Facebook has been charged with housing discrimination by the US government." *The Verge* (March 28, 2019) Retrieved from: https://www.theverge.com/2019/3/28/18285178/facebook-hud-lawsuit-fair-housing-discrimination.

31. Zuboff, *The age of Surveillance Capitalism.*

32. Eubanks, *Automating Inequality.*

33. Dillon, G. L. "Introduction: Indigenous futurisms, Bimaashi Biidaas moose, flying and walking towards you." In *Walking the clouds: An anthology of indigenous Science Fiction,* edited by G. L. Dillon Tucson, AZ: University of Arizona Press.

34. Dillon, G. L. 2012. "Introduction: Indigenous futurisms, Bimaashi Biidaas mose, flying and walking towards you." In *Walking the clouds: An anthology of indigenous Science Fiction,* edited by G. L. Dillon Tucson, AZ: University of Arizona Press.

35. Madsen, D. L. "The mechanics of survivance in indigenously-determined video-games: Invaders and never alone." *Transmotion,* 3(2) (2017), 79–110.

36. Ibid.

37. LaPensée, E. (B. A. Lameman) and J. E. Lewis (2011) "Skins: Designing Games with First Nations Youth," *Journal of Game Design & Development Education* 1(1), Retrieved from: www.rit.edu/gccis/gameeducationjournal/skins-designing-games-first-nations-youth.

38. Lewis, J., and S. T. Fragnito. "Aboriginal territories in cyberspace." *Cultural Survival Quarterly,* 29(2) 2005), 29.

39. Butler, O. E. *Parable of the talents: A novel* (Vol. 2). Seven Stories Press (1998).

40. Robinson, M. R., and C. Neumann. "Introduction: On Coogler and Cole's Black Panther film." *Global perspectives, reflections and contexts for educators. Journal of Pan African Studies,* 11(9) (2018), 1–13.

41. Eglash, R., M. Krishnamoorthy, J. Sanchez, and A. Woodbridge. "Fractal simulations of African design in pre-college computing education." *ACM Transactions on Computing Education (TOCE),* 11(3) (2011), 1–14.

42. Sheridan, K. M., K. Clark, and A. Williams. "Designing games, designing roles: A study of youth agency in an urban informal education program." *Urban Education,* 48(5) (2011), 734–758.

BIBLIOGRAPHY

Brandom, R. 2019. "Facebook has been charged with housing discrimination by the US government." *The Verge* . Retrieved from: https://www.theverge.com/2019/3/28/18285178/facebook-hud-lawsuit-fair-housing-discrimination

Buolamwini, Joy. 2019. "Actionable auditing: Coordinated Bias Disclosure Study." *MIT Media Lab.* Retrieved from: https://www.media.mit.edu/projects/actionable-auditing-coordinated-bias-disclosure-study/overview/

Butler, O. E. 1998. *Parable of the talents: A novel* (Vol. 2). Seven Stories Press.

Condis, M. 2015. "No homosexuals in star wars? BioWare, 'gamer' identity, and the politics of privilege in a convergence culture." *Convergence*, 21(2), 198–212.

Crooks, H. R., and S. Magnet. 2018. "Contests for meaning: Ableist rhetoric in video games backlash culture." *Disability Studies Quarterly*, 38(4) Retrieved from: https://dsq-sds.org/article/view/5991/5142.

Cupples, J., and R. Grosfoguel. (eds.). 2018. *Unsettling eurocentrism in the westernized university*. Routledge.

Dastin, Jeffrey. 2018. Amazon scraps secret AI recruiting tool that showed bias against women. *Reuters*. Retrieved from: https://www.reuters.com/article/us-amazon-com-jobs-automation-insight/amazonscraps-secret-ai-recruiting-tool-that-showed-bias-against-women-idUSKCN1MK08G

Dillon, G. L. 2012. "Introduction: Indigenous futurisms, Bimaashi Biidaas mose, flying and walking towards you." In *Walking the clouds: An anthology of indigenous Science Fiction*, edited by G. L. Dillon Tucson, AZ: University of Arizona Press.

Eglash, R., M. Krishnamoorthy, J. Sanchez, and A. Woodbridge. 2011. "Fractal simulations of African design in pre-college computing education." *ACM Transactions on Computing Education (TOCE)*, 11(3), 1–14.

Eubanks, Virginia. 2018. *Automating Inequality: How High-Tech Tools Profile, Police, and Punish the Poor*. St. Martin's Press.

Gee, B., & Peck, D. "The illusion of Asian success: Scant progress for minorities in cracking the glass ceiling from 2007–2015". *Ascend* (2019). Retrieved from: https://cdn.ymaws.com/www.ascendleadership.org/resource/resmgr/research/theillusionofasiansuccess.pdf

Gershon, I., and J. S. Taylor. 2008. "Introduction to 'in focus: Culture in the spaces of no culture." *American Anthropologist*, 110(4), 417–421. doi: 10.1111/j.1548-1433.2008.00074.x

Gooch, D., M. Barker, L. Hudson, R. Kelly, G. Kortuem, J. V. D. Linden, . . . J. Mackinnon. 2018. "Amplifying quiet voices: Challenges and opportunities for participatory design at an urban scale." *ACM Transactions on Computer-Human Interaction (TOCHI)*, 25(1), 1–34.

Habermas, J. 2015. *The lure of technocracy*. Cambridge, UK and Malden, MA: Polity

Hegewisch, A., C. Childers, and H. Hartmann. 2019. "*Women, automation, and the future of work*." Institute for Women's Policy Research. Retrieved from: https://iwpr.org/publications/women-automation-future-of-work/

Hao, K. 2019. "The future of AI research is in Africa." *MIT Technology Review*. Retrieved from: https://www.technologyreview.com/2019/06/21/134820/ai-africa-machine-learning-ibm-google/

LaPensée, E. (B. A. Lameman) and J. E. Lewis (2011) "Skins: Designing Games with First Nations Youth," Journal of Game Design & Development Education 1(1), retrieved from www.rit.edu/gccis/gameeducationjournal/skins-designing-games-first-nations-youth.

Lee, T. B. 2019. "Richard Stallman leaves MIT after controversial remarks on rape." *Arts Technica*. Retrieved from: https://arstechnica.com/tech-policy/2019/09/richard-stallman-leaves-mit-after-controversial-remarks-on-rape/

Leonard, D. J. 2006. "Not a hater, just keepin'it real: The importance of race-and gender-based game studies." *Games and Culture*, 1(1), 83–88.

Lewis, J., and S. T. Fragnito. 2005. "Aboriginal territories in cyberspace." *Cultural Survival Quarterly*, 29(2), 29.

Lorde, Audre. (2018) *The Master's Tools Will Never Dismantle the Master's House. Penguin Classics.*

Madsen, D. L. 2017. "The mechanics of survivance in indigenously-determined video-games: Invaders and never alone." *Transmotion*, 3(2), 79–110.

Woetzel, J. Manyika, J., Dobbs, R.,Madgavkar, A., Ellingrud, K., Labaye, E., Devillard, S., Kutcher, E, Krishnan, M. 2015. "How advancing women's equality can add $12 trillion to global growth." McKinsey & Company.Retrieved from: https://www.mckinsey.com/featured-insights/employment-and-growth/how-advancing-womens-equality-can-add-12-trillion-to-global-growth

Murdock, J. 2020. "Damien Patton, CEO of AI surveillance tool Banjo, resigns after KKK Neo-Nazi past exposed." *Newsweek*. Retrieved from: https://www.newsweek.com/banjo-ceo-damien-patton-resigns-nazi-white-supremacy-ai-social-media-mining-police-1503339

Noble, S. U. 2018. *Algorithms of oppression: How Search Engines Reinforce Racism.* New York: New York University Press.

Page, S. 2017. "Just having people who look different isn't enough to create a diverse team." *LinkedIn*. Retrieved from: https://www.linkedin.com/pulse/just-having-people-who-look-different-isnt-enough-create-scott-page/?published=t

Pitchbook. 2019. "PitchBook Universities." Retrieved from: https://pitchbook.com/news/articles/pitchbook-universities-2019

Prasad, A., and A. Mills. 2010. "Critical management studies and business ethics: A synthesis and three research trajectories for the coming decade." *Journal of Business Ethics*, 94, 227–237.

Robinson, M. R., and C. Neumann. 2018."Introduction: On Coogler and Cole's Black Panther Film (2018): Global Perspectives, Reflections and Contexts for Educators.". *Journal of Pan African Studies*, 11(9), 1–13.

Rosenberg, S. 2017. "Silicon valley's first founder was its worst." *Wired*. Retrieved from: https://www.wired.com/story/silicon-valleys-first-founder-was-its-worst/

Saleem, M., and C. A. Anderson. 2013. "Arabs as terrorists: Effects of stereotypes within violent contexts on attitudes, perceptions, and affect." *Psychology of Violence*, 3(1), 84–99.

Saunders, A. 2016. "Blogs. Inclusive innovation incubator." Retrieved from: https://www.in3dc.com/blog-1

Sheridan, K. M., K. Clark, and Williams. 2013. "A. Designing games, designing roles: A study of youth agency in an urban informal education program." *Urban Education*, 48(5), 734–758.

Šisler, V. 2008. "Digital Arabs: Representation in video games." *European Journal of Cultural Studies*, 11(2), 203–220.

Tae-Hee, J. 2016. "Frederic S. Lee and his fight for the future of heterodox economics." *PSL Quarterly Review*, 69(278), 267–278.

Vermeulen, L., M. V. Abeele, and S. Van Bauwel. 2016. "A gendered identity debate in digital game culture." *Press Start*, 3(1), 1–16.

Vu, V., C. Lamb, and A. Zafar, A. 2019. "Who are Canada's tech workers?" *Brookfield Institute for Innovation + Entrepreneurship*. Retrieved on August 30, 2019, from: https://brookfieldinstitute.ca/report/who-are-canadas-tech-workers/

Weststar, J. 2015. "Understanding video game developers as an occupational community." *Information, Communication and Society* 18(10), 1238–1252.

Witte, A. E. 2012. "Making the case for a post-national cultural analysis of organizations." *Journal of Management Inquiry* 21(2), 141–159.

World Economic Forum. 2018. *The Global Gender Gap Report*. Retrieved from https://www.weforum.org/reports/the-global-gender-gap-report-2018

Zuboff, S. 2019. *The age of Surveillance Capitalism: The fight for a Human Future at the new frontier of power*. Profile Books.

Chapter 4

"The Specter of Self-Organization"

Will Algorithms Guide Us toward Truth?

Ravi Sekhar Chakraborty

The idea of self-organization has an undeniable virality in contemporary conversations about algorithm-driven technologies. As often happens, such a currency makes a concept ripe for appropriation in ever newer contexts until it acquires the aura of a new metaphysics. It is then an occasion to step back and historicize the concept so the true novelty of its latest incarnation is understood. Way before the prevalence of an algorithmically determined existence, the maverick French thinker Gilles Châtelet anticipated the desire for control, which underlies the promise of self-organization. Châtelet's prophecy sounds less radical today. Here is a glimpse of it:

> A science, the general theory of networks and systems-cybernetics—will offer its services, permitting audacious "social engineers" to push back the frontiers of methodological individualism—to conceive scenarios that, not too long ago, no average man would have dared dream of: to transform thermocracy into neurocracy, to succeed in fabricating behaviors that will guarantee a watertight barrier against political intelligence.[1]

Gilles Châtelet's notion of neurocracy refers to a manifestation of the kind of algorithmic culture that ignores the problem of manipulating public opinion for the promise of managing private belief. As a seemingly spontaneous aggregate of private beliefs, it provides potentiality for a politics that is robbed of voluntarism and solidarity. The consequent question that arises is whether this chaotic aggregate can be engineered. If so, is this engineering only the latest step in the historical project which began with constructing the fiction of a free individual in a market democracy? The question of the spontaneous emergence of truth in an algorithmic culture of control is of preeminent importance here.

The metaphysical, rather than the spiritual, echoes of the question of truth are intended with only some degree of irony so that it is easier to confront the epistemological reality where algorithms control so much of our access to knowledge. Many dimensions of contemporary culture are governed now largely by algorithms. Their recommendations, classification, and other modes of organization govern everything ranging from political opinion to product choices, from taste in music to the field of medicine. It is as if in this new ontology, the question of truth comes with a theological spin that is only a matter of exaggeration. The pervasive reach of algorithms asks us whether one should rely on them just to make for a faster and more efficient decision about choice or one can rather push the envelope further and ask the fundamental question of truth. Truth signifies then the consequence of the teleological turn that is expectedly attributed to any hegemonic discourse, or in this case, a discursive environment driven by algorithms.

TECHNOLOGICAL CULTURE/ CULTURAL TECHNOLOGY

Before we set out to explore the relationship between choice, truth, and algorithmic logics, it is important to clarify what this chapter will steer clear of: the conversation about algorithmic culture that is usually marked with cynicism about privacy and regret over being manipulated. It is true that the algorithms controlling our lives often affirm the "average" or the "crowd" perspective in the name of customization. Algorithms now raise important questions about moral responsibility (i.e., how they direct self-driving cars into accidents and how they recommend divisive political advertisements). Nonetheless, an apocalyptic slant to such discussions often misrepresents a nuanced view of algorithmic culture and discounts a properly critical perspective. This simplistic paranoia about algorithmic culture tends to endorse a Luddite ring, which makes it even less credible. How can we possibly discuss algorithmic culture so that we sound not so much blindsided by the contemporary condition but see it along the expected lines of evolution that is not defined entirely as technological? The spirit of Gilbert Simondon's thought provides a cue for our own purposes. For Simondon, the relationship between technology and society is not to be thought of as two individuated entities interacting with each other. Technology is always caught at various stages of individuation in society and vice versa. Both the cyberutopians and the doomsayers are party to a sustained indifference to the actual process of psycho-social individuation. Algorithmic culture should be situated in a matrix of the most divergent set of discourses: from popular conceptions to specialist's insights, from research themes in sciences such as theoretical biology, and from computer science and economics to corporate-speak on algorithmic utopias.

Here the theme of "algorithmic culture" needs to be put on a firmer foundation by historicizing it and distinguishing it from its antecedents. Does algorithmic culture imply a culture of technological design? Or is it meant to connote the cultural effects of various technological approaches? Or do algorithms provide a general style of thought that is meant to organize actions beyond the sphere of technology? Slack and Wise develop a valuable distinction between "cultures of technology" and the cultural effects of technology.[2] It is essential to see that these conceptions coexist with one another. Such a conceptual atmosphere is best suited to investigate the concerns of the present chapter for two broad reasons. First, the social physics view of human culture encourages us to think of yet another possibility: culture *as* technology. Further, a technology that facilitates certain kinds of interaction between its constituents which are individual human beings. Second, the specificities of the history of algorithms both shape and are shaped by how algorithms and their role as perceived emerge as a culture of technology. The central argument of this chapter is to show a formal continuity between the evolution of culture *as* technology and the emergence of the culture of algorithmic technologies. This will be done through tracing the history of the theme of self-organization that animates contemporary visions of algorithmic utopia.

While this chapter does not engage a particular case study, it draws from the philosophy of algorithm design and implementation in order to understand the broader contours of the phenomenon. This theoretical gesture marks a shift beyond studies framed as the "social and cultural impact of technology" which seemingly de-emphasizes or even rejects the importance of any domain knowledge of the technologies involved. We must not presume a pre-given algorithmic culture whose effects we are setting out to assess. This chapter is an attempt to define an algorithmic culture in the first place.

Algorithms in themselves suggest only an adherence to step-by-step procedure, as Donald Knuth in his iconic *Art of Computer Programming* suggests, toward problem-solving.[3] Knuth states that the working definition of algorithm demands finitude to the number of steps it takes. Such a set of instructions where each instruction depends on the previous one becomes synonymous with efficiency much later when punch cards are employed in mechanical computing devices. If one goes by the history of algorithms in Arabic, one can say that algorithms were part of a culture of mathematics long before they were conceived as a cultural foundation on their own right. Yet it can be asked if there was an aura attributed to algorithms that defined their use in that culture. Between medieval abacisists and "algorists," the work of the algorist clearly represented a cleverer, less laborious manner to do certain kinds of calculations.

With the introduction of punch cards to improve the speed of computations, eventually, the definition of algorithms gives way to manufacturing processes controlled by the computer. We were still not perceiving of algorithms as being given the sort of autonomy to have a life of their own and to be entrusted to execute regimes of self-organization. Algorithms were still part of the cybernetic machinery and had not yet been liberated from the body in the technological imagination. In his speech titled "Men, Machines and the World about Them," Norbert Wiener,[4] the father of cybernetics, talks of the possibility of a production system that can be adapted to different designs by changing the "taping," that is, changing the algorithm for the design while keeping the same set of physical machineries. As such, an advance in efficiency and versatility of the factory is seen as an extension of the cybernetic improvement on factories that are more manually governed. We note that the ideological regime of control still remains.

The lesson to be drawn from the cybernetic interpretation of algorithms is that one can make a compelling case for algorithms as being the centerpiece of a cultural technology that is installed as an evolution on preceding technologies. Asking about the "social implications" of algorithms is potentially fallacious because algorithms are, to emphasize again, not isolated abstractions empowered with meaningful agency to affect their own political implications. Here one should not be making the trivial assertion that algorithms are social merely because they are designed and implemented within a society. As a matter of form and structure, markets, democracies, and algorithm-driven technologies are all part of cultural technologies accumulating their own histories of evolution embodying their most contemporary manifestations. The message of this insight is taken a step ahead if we notice how the very science, nay, mathematics of algorithm design is deeply inflected by what may be dismissed as superficial prejudices about what technologies can do. This aspect will be underlined in the next section.

ALGORITHM: AN UNCERTAIN MACHINE

Let us, for a moment, pause and consider a truly internal history of technology to defamiliarize ourselves with the prevailing association of algorithmic culture with control. A sincere focus on the history of algorithms will enable us to tease out the influence of algorithms that cannot be subsumed within rightfully critical discussions of the effects of the Internet. A humble ambition of this chapter is to enrich, if not be distracted by the skepticism about the Net Delusion as articulated by Evegeny Morozov, among others.[5] Such a perspective does not necessarily accord algorithms any decidedly deterministic role in leading to the current technological condition. However, if we were to take up the mantle of speculating enough to substantiate the ideologies embedded in the design and implementation of algorithms, it would be possible to construct a narrative beyond easy presentism centered around control and manipulation, especially with respect to what algorithms do today in their most pervasive incarnations in the techno-capitalist imaginary.

It turns out that, as a culture, algorithms have historically not always been associated with control but rather with uncertainty. Here, when we speak of algorithms, we specifically mean sets of instructions implementable on a computer. When Alan Turing conceived of the abstract machine that could implement all kinds of computation, he inaugurated fields of possibility whose scope was mathematically difficult to grasp. In his classic paper, titled "Computational Machinery and Intelligence," Turing writes:

> The view that machines cannot give rise to surprises is due, I believe, to a fallacy to which philosophers and mathematicians are particularly subject. This is the assumption that as soon as a fact is presented to a mind all consequences of that fact spring into the mind simultaneously with it. It is a very useful assumption under many circumstances, but one too easily forgets that it is false.[6]

Algorithms being associated with control and certainty are not just how technology is perceived by nontechnologists, but it is a definitive centerpiece of the algorithm designer's worldview. How is it that it was not understood that the limits of what algorithms can do is necessary to substantiate how much we attribute a sense of "control" to algorithmically driven activities? Today, the field of theoretical computer science devotes itself to studying the limits of algorithms. With the increase in computing power, it is possible to repeat certain algorithmic steps enough number of times that certain seemingly impossible tasks can potentially be completed (or not!). The mere repeatability of the steps of an algorithm does not entail a certain knowledge of the result of the running of the algorithm even potentially infinite number of times.

What Turing's prescient observation shows is that there is a dialectic between control and uncertainty that unwinds over the succeeding century in the way that algorithms furnish not just a technology but an ideology, nay, a culture of control. This chapter intends to tease out the dynamic interplay of the mentalities of control and uncertainty, which contributes to the expectation of self-organization. The implementation of algorithms on computers, while making certain computations less time-consuming, also made us revise fundamentally any notion of certainty associated with the power of algorithms.

Shall we say that self-organization provides a solution to the problem of the dialectic between control and uncertainty? Attempting such a reading of the tropology of self-organization in the discourse of algorithmic technologies is inspired by what can be called a mathematical tropology pioneered by the French philosopher of mathematics, Albert Lautman. The discourse of this technology and the discourse of mathematics can very well be animated by similar dialectics. For Albert Lautman, mathematical theories represent solutions to dialectics of Platonic ideas.[7] For example, he interprets the real number line to be representing one between the notions of continuity and discontinuity. Why can't one apply this to think of the discourse of algorithmic technologies as negotiating another dialectical pair? I reiterate my conjecture again: self-organization is the formal solution whose condition of existence is the dialectic of control and uncertainty.

GILES CHÂTELET: AN UNTIMELY THINKER OF THERMOCRACY

The most apt manner of historicizing what can be called an algorithmic politics is a development of the thought of French philosopher Gilles Châtelet. Known essentially as an epistemologist of science and mathematics, his polemic commentary on contemporary politics, *To think and Live like Pigs*, was published in 1998 predating mainstream discussions of algorithmic politics.[8] Châtelet's creative interpretation of the "social physics" of market democracies bears the refreshing influence of his preoccupation with the history and philosophy of the sciences. If one were to follow this genre of analysis, then we are too late already in responding to algorithmic culture. It would not be preposterous for Châtelet to extract a philosophy from algorithmic thought that would be useful for thinking about social phenomena that might be quite anachronistic with respect to the contemporary rise of algorithms.

A proper tribute to Gilles Châatelet is not without an adequately polemical dash to the argument. Châtelet, preceding Morozov, begins where Morozov's thought seems to end. Morozov, writing on the "Net Delusion," was one of the most prominent, if not the first, to sound the ethical conundrum on naive optimism about the Internet and its possibilities.[9] He was ahead of others in his skepticism of technological solutionism. But despite his most astute observations, Morozov sounds like any other detractor of technological determinism and its so-called ethical implications. Given how Châtelet has faded into relative obscurity, it is even more indispensable to observe the value of his prescience.

Châtelet was able to foresee the Morozovian interpretation much before the Internet became the pervasive reality that it is today. There is no trite one-upmanship being introduced with regard to competing astrologies about the future of technology. But the failure to predict is actually a reflection of the failure to properly historicize or be ignorant of it. A radical historicization of the sort pursued by Châtelet refused neat considerations of the cause and effects of supposedly new technologies. Social physics provides almost a formalist's re-examination of what exactly makes a new technology formally new and with respect to which formally defined contexts.

Notice again the salience of the phrase "the social physics of the market." For Châtelet, such social physics amounts inevitably to social thermodynamics. What is the most iconic connection we make with concepts of temperature introduced in high school? We picture boxes of gas molecules buzzing about in high speed and temperature being an estimate of their average kinetic energy. It is a moot question whether the history of thermodynamics as a subject literally influenced the political imagination. But we can at least say that the free individual in emerging market democracies is not much unlike the free molecule with certain degrees of freedom. Be it the context of the state or capitalism, the epistemological problem of understanding the direction of popular consciousness was like measuring the temperature of a mass of fast-changing opinions.

We must remember here that notions of thermodynamics were not alien to the sociological imaginary. Levi Strauss's use of the notion of entropy in *Tristes Tropiques* shows that there is certainly an interest in disorder as much as there is in the question of structure.[10] This concept is further fleshed out by the work of Mauro Almeida.[11] Levi Strauss's interest in disorder is in stark contrast to the problem of chaos or permanent disorder that is to be tackled by social thermodynamics proper. But Levi Strauss's view would not be called a social physics because there is no implicit assumption about molecular components like individual actions being the fundamental building blocks of society. Hence, such an anthropological view is not a good candidate for building a continuity with the concept of thermocracy.

Instead, we may consider a perusal of the thoughts of a scientist who truly championed self-organization as a physical principle whose fundamental nature we cannot afford to ignore. We are speaking of the Nobel Prize winning physicist, Ilya Prigogine. Interestingly, Prigogine has well-articulated remarks to make about what he calls the "networked society" which we may take to be a corollary of algorithmic culture.[12] Specifically, he talks of the emergence of complex network societies in the following manner:

> First, I feel that there is some analogy between the present evolution toward the networked society and the processes of self-organization I have studied in physics and chemistry. Indeed, nobody has planned the networked society and the information explosion. It is a remarkable example of spontaneous emergence of new forms of society. Complexity is moreover the key feature of far-from-equilibrium structures. The networked society is of course a non-equilibrium

> structure which emerged as a result of the recent developments in Information Technology.[13]

Furthermore, Prigogine displays a keen awareness about the inequalities the networked societies may bring *despite* their complexity. It is important to hear his prescient voice directly:

> I believe that in the future the networked society will be judged according to its impact on the inequality between the nations. Of course, there are well-known advantages to the networked society. . . . Will the networked society be a step in the direction of the realization of this goal? From this point of view, it is interesting that each bifurcation in the past resulted in people who benefited from it and in people who became victims. The neolithic revolution led to extraordinary gains in the field of the arts. It led to the construction of pyramids for the pharaohs but also to common graves for the common people. However, slavery probably also started with the neolithic civilization and has continued up to this day. Similarly, the industrial civilization led to the development of the proletariat in addition to increased wealth.[14]

We will not let go of the nuanced distinction between algorithmic culture and network society, but Prigogine's structural view helps to close the gap between culture as technology and a technological culture, which is the goal of this chapter. Armed with Prigogine's faith in the emergence of complex structures and his skepticism about the benefits they promise, we return to Châtelet's imagination of thermocracies. Thermocracies demonstrate that chaos has not died down after we left the state of nature. In the new marketplace of opinions and beliefs, the chaos is heating up again, almost about to tip over into something very ominous, and yet it does not. The frenzy of political opinions has replaced the politics of the streets. Thermodynamically speaking, there has been a redistribution of political energy, and nothing about it is inherently detrimental to the idea of politics. The factory required us to envision a containment of political energy within its boundaries. The market released the energy until it got dissipated in service of the invisible hand. There is a massive effort to regain this dissipated energy, but not in its original form.

It is here that Châtelet speaks of the rise of cybernetic networks and their eventual destiny in ushering the evolution of thermocracies into neurocracies. For then, it is no longer necessary to invent the "average man" just like we find "average kinetic energy" when we calculate the temperature. Because it is now possible to manipulate decisions and choices as finely defined as a mouse click or eye-gaze, the neurocracy seemingly warrants a much higher degree of control and yet there is a bargain because such an extreme decentralization of control brings back a degree of uncertainty about the relevant totality. The network is global yet this global form is an emergent totality, not a planned one. Neurocracies do not need averages; they work with each unique individual as a personalized molecular unit. There may be incidental monopolies in how neurocracies are governed by states or large corporations, but the cybernetic network by the very virtue of its structure is not amenable to the spirit of total central control. Thermocracies were still actively concerned with totalities: totalities of citizen or consumer population and the very optimistic about the controlling of such totalities. Neurocratic totalities are elusive and unmindful of boundaries of communities, nations, or other kinds of groups. All such boundaries are translatable into variables and parameters and thus are subjected to algorithmic design.

To continue the metaphor further, algorithms are means for extracting work out of the heat of social chaos understood in this manner. But algorithms are not engines. Algorithms do not guarantee any rate of efficiency for reaching any well-defined goals. Algorithms are varied and may be at cross-purposes with each other. Algorithms, instead, aspire to govern the micro-decisions of molecular individuals. There is an equilibrium that has made mass voluntarism in politics irrelevant. It is important to underline that the very idea of the market created the conditions of possibility of neurocracy. Nevertheless, algorithms furthered the scope of the kind of choice given up at the altar of the market. Earlier, such choices were limited to our roles in the market, but now all behavior is potentially in the ambit of a marketplace of beliefs and attitudes. Dating is modeled as an auction. This marketplace is developing its own terms of self-organization and concomitant myths.

But it was not as if mythologies were already being designed to cope with the chaos breeding within thermocracies. Short of catastrophe, the survival of the market is a mark of the success of the idea of the markets or so we want it to be. The "invisible hand," as conceived by Adam Smith in the *Wealth of Nations*, was meant to take care of precisely this kind of scenario.[15] Jean-Pierre Dupuy astutely points out that there is a consistency to be discovered between the ideas explored in "Theory of Moral Sentiment" and "The Wealth of Nations."[16] To develop the conditions for the possibility of the invisible hand, Smith is devoted to creating a complete theory of human nature and behavior. Dupuy goes on to criticize economists for oversimplifying this aspect of Smith's thought and building self-absorbed models which calculate interactions between individuals supposed to be self-interested. Intersubjectivity is ignored, and a very malnourished notion of the subject remains. At the level of the newest theoretical culture of algorithmic game theory, the algorithmization of the problem of finding the market equilibrium also dissolves what Dupuy calls the "intersubjective" element of the interactions in the market.

If we were to trace the algorithmic condition right up to its underlying presumptions, we notice an implicit theory of human nature, no less, emerging there too. But this theory does away with the idea of the subject and replaces it with the human brain, nay, brain compartments and neurotransmitters. A neurocracy attempts to individuate not just the individual but also the constituent brain activities which define him/her. The algorithm interacts not with the subject but with the gaze of the eye and the click of the mouse, all of which can be reduced to neural activity. Much as it has been pointed out that it is a fallacy to hope that the brain can be reduced to its neural activity, it has not, however, stopped dreams of rewiring the world into a neurocracy of convenient disorder rather than betting on a thermocratic pre-cybernetic illusion of order.

Much as the myth of the invisible hand was conjured up to make the chaos of the market look benign, Châtelet asks if the "chaos of opinions and micro-decisions will always lead to something reasonable."[17] This chaotized aggregate of opinions, according to Châtelet, always leads to a predictable average that can be politically contained. He calls it a "thermocracy," which allows for "a seductive market-chaos of opinions as a 'natural' parameter or thermometer."[18] Most pertinently, he seems to anticipate the digital cyber context that is uniquely associated with the Post-Truth era. He is the first to speak of the emergence of a "neurocracy" from this thermocracy where the unit of control shall be narrowed further from the will of the atomized individual to the smaller set of her/his many micro-decisions and beliefs. The possibilities of such control are afforded by the cybernetic existence of the Internet era, as foreseen by Châtelet.

THE RECURRING NIGHTMARE OF CHAOS

Gilles Châtelet offers the illuminating perspective of juxtaposing individual freedom against the histories of other individual freedoms, real and theoretical. We may hark back to the English philosopher Thomas Hobbes. In *Leviathan*, Hobbes conceives of the state of nature as one where individual freedom has brutal consequences for the society at large because of the violence and pillage that is bound to ensue.[19] Hence, this is one example of how one articulates the notion of freedom. Other notions could include the freedom to buy and sell in the market. The notion of absolute freedom in the Hobbesian state of nature may lead to absolute chaos. Theoretically, both economic freedom in the market and political freedom in democracy represent a bargain out of Hobbesian chaos. Such formal guarantees of liberty are still synonymous with an embrace of various degrees of chaos. As pointed earlier, all these pictures assume a certain molecular vision of society where all these molecular individuals represent a population that awaits a passage from disorder to order, to put it in thermodynamic terms. The poignancy of this liminal position where the market and the democracy become indistinguishable is properly established if we acknowledge its situation between the fictions of Hobbesian chaos and near-utopian self-organization.

A nod to Levi Strauss cannot be avoided here. Levi Strauss eschewed the choice of chaos as the central paradigm of studying any community. Being primarily invested in the idea of structure, Levi Strauss is keen about the possibility of disorder only as a matter of change in structure. Why do certain structures degenerate into disorder? This explains Levi Strauss's preoccupation with the concept of entropy in *Tristes Tropiques*.[20] The thermodynamic notion of entropy is a measure of the chaotic behavior of molecules. Levi Strauss hints toward this possibility of an operating paradigm of chaos without exactly adopting it. A social physics is as vulnerable to the limitations and advantages of choosing its own paradigm as is any other science. And we must remember it is not even a science proper. It is a paradigm that forms less the foundation of a discourse of knowledge but instead that of a discourse in action. A philosophical archaeology of such a discourse is to be done with due attention to the expansive scope of its genealogy.

There are clear antecedents which help to comprehend this vision of chaos brought about by teeming individual freedoms jostling with each other. Châtelet contrasts the benign chaos of the free market with the destructive chaos of the Hobbesian state of nature to characterize the illusion of freedom that defines market democracies. This comparison allows him to comprehend the chaos of opinions that is being staged in the Post-Truth context. Any chaos comes with the promise of self-organization as well as the anxiety of anarchy. The uneasy stability of the opinion market is the only possible bargain of an equilibrium.

Algorithms offer an imaginary where chaos can be contained, or rather "optimized." One can begin from field of Operations Research, which is a legacy of scientific management techniques applied in contexts as varying as flight scheduling to factory task allocation. Algorithms have been, for a long time, a direct corollary of humanity's growing skill in industrial levels of organization. But with the advent of information theory, algorithms are implicated in the ushering in a new metaphysics itself which is the metaphysics of information. Everything is information, and algorithms are the best instruments to capture the information that is relevant in any context. Most importantly, algorithms are now being employed to unscramble the underlying mechanisms of seemingly spontaneous complex processes such as evolution and market equilibrium. Leslie Valiant has passionately argued for an algorithmic basis to Darwinian evolutionary theory which he says is insufficient to explain how simple organisms evolve into complex ones.[21] Within economics, Vijay Vazirani has written about algorithms being the natural candidate to prove the benevolence of the invisible hand whose mathematical ratification, though beginning around half a century ago, is only now benefiting from advances in computer science.[22]

VISIONS OF EQUILIBRIUM

The theme of the market equilibrium and efforts to compute it have dominated research in economics, extending its influence into computer science. Such a question of the possibility of a spontaneous equilibrium or spontaneously arrived design is also the guiding metaphor in studies of other complex systems such as biological sciences, specifically questions like how life emerges out of inanimate matter, biological growth, among others. Even within research culture, there is a convergence of economics and computer science which is directed toward computing and finding an equilibrium point and not being satisfied with the theological sleight of hand that is the "Invisible Hand." In other words, the promise, or even proof of the existence of the invisible hand may not be enough to guarantee an equilibrium.

That is precisely the conclusion of Constantinos Daskalakis, the winner of the 2018 Rolf Nevalinna prize in computer science. He concluded that it would take prohibitively long to reach the equilibrium at least in scenarios idealized in terms of game-theoretic assumptions. Even as computer scientists like Daskalskis attempt to find methods to compute the equilibrium, it is the specter of the equilibrium or the hope for self-organization that still guides not just a faith in markets but also the algorithms driving the market.[23] Algorithms have been enlisted to theoretically secure the utopian teleology underlying market discourse even as actual algorithms quickly determine our own "neurocratic" becoming in the marketplace. Such a mutation of a vested interest in control to risking confidence in self-organization is basically the corollary of factory-driven thermocracies morphing into algorithm-driven neurocracies. Hence, the work of Daskalskis, which is at the intersection of economics and computer science, is representative of the cross-disciplinary perfusion of self-organization as an epistemological trope.

The algorithmic culture of control can then be seen as only furthering a schematic of society-individual relationship which has its roots in the theory of the market democracy. Algorithmic culture is not about surveillance and control in the totalitarian sense but about attaining efficiency in staging a kind of chaos. It is essential to isolate the question of morally evaluating how well the chaos self-organizes or whether self-organization is even necessary. The ethical concerns are not impotent afterthoughts either. However, the historicizing of a phenomenon by laying bare its evolution within a matrix of ideas and inventions cannot be blinded by the overwhelming judgment of associated values.

Self-organization is then the destined guiding trope of algorithmic culture. The theme of self-organization dominating the algorithmic worldview often sees its most caricatured exaggeration in the nightmare of the superhumanly intelligent machine that will be a threat to the human race. Such a projection into the future is perhaps a reflection of how much algorithms have already been entrusted to explain *and* define complex and apparently spontaneous processes. The superhuman piece of artificial intelligence makes the algorithmic theology complete but it is not what defines it. Self-organization is the rea teleological impulse organizing algorithmic culture.

NOTES

1. Châtelet, Gilles. *To Think and Live Like Pigs*. Translated by Robin Mackay. Sequence Press, 2014 (Originally published in 1998).
2. Slack, J., and J. M. Wise. *Culture Technology: a Primer*. Peter Lang, 2007.
3. Knuth, Donald Ervin. *The Art of Computer Programming*. Addison-Wesley, 2010.
4. Wiener, Norbert. "Men, machines, and the world about them." *WNYC*, October 25, 1950.
5. Morozov, Evgeny. *The Net Delusion: How Not to Liberate the World*. Penguin, 2012.
6. Turing, A. M. "Computing machinery and intelligence." *Mind*, vol. 59, 1950, p. 46.
7. Lautman, Albert, and Simon B. Duffy. "Mathematics, ideas and the physical real." *Continuum*, 2011, 197–220.
8. Châtelet Gilles. *To Live and Think like Pigs: The Incitement of Envy and Boredom in Market Democracies*. Translated by Robin Mackay. Sequence Press, 2014 (Originally published in 1998).
9. Morozov. *The Net Delusion*.
10. Lévi-Strauss Claude. *Tristes Tropiques*. Translated by John Weightman and Doreen Weightman. Penguin Books, 2012.
11. Mauro W. Barbosa de Almeida, et al. "Symmetry and entropy: Mathematical metaphors in the work of Levi-Strauss [and Comments and Reply]." *Current Anthropology*, vol. 31, no. 4, 1990, pp. 367–385. *JSTOR*, www.jstor.org/stable /2743257. Accessed May 2, 2020.
12. Prigogine, Ilya. "The Networked Society." *Journal of World-Systems Research*, 2000, Vol. 6. no 3. pp. 892–898.
13. Prigogine, "The Networked Society," p. 893.
14. Ibid., p. 895.
15. Smith, Adam. *Wealth of Nations*. Vintage, 2020.
16. Dupuy, Jean-Pierre. "Invidious Sympathy in the Theory of Moral Sentiments." *Revue du MAUSS*, vol. 31, no. 1, 2008, pp. 81–112.

17. Châtelet, *To Live and Think like Pigs*, p. 24.
18. Ibid., p. 60.
19. Hobbes, Thomas. *Leviathan*. Ancient Wisdom Publication, 2019.
20. Lévi-Strauss, Claude. *Tristes Tropiques.*
21. Valiant, Leslie G. "Evolvability." *Journal of the ACM (JACM)*, vol. 56, no. 1, 2009, pp. 1–21.
22. Vazirani, V. V. "Can the theory of algorithms ratify the 'invisible hand of the market'?" In Hirsch E.A., Karhumäki J., Lepistö A., Prilutskii M. (eds), *Computer Science—Theory and Applications*. CSR, 2012. Lecture Notes in Computer Science, vol. 7353. Springer, Berlin, Heidelberg.
23. Daskalakis, Constantinos, Paul W. Goldberg, and Christos H. Papadimitriou. "The complexity of computing a Nash equilibrium." *SIAM Journal on Computing*, vol. 39, no. 1, 2009, pp. 195–259.

BIBLIOGRAPHY

Châtelet Gilles. *To Live and Think Like Pigs: the Incitement of Envy and Boredom in Market Democracies*. Translated by Robin Mackay. Sequence Press, 2014 (Originally published in 1998).

Daskalakis, Constantinos, Paul W. Goldberg, and Christos H. Papadimitriou. "The complexity of computing a Nash equilibrium." *SIAM Journal on Computing*, vol. 39, no. 1, 2009, pp. 195–259.

Dupuy, Jean-Pierre. "Invidious sympathy in the theory of moral sentiments." *Revue du MAUSS*, vol. 31, no. 1, 2008, pp. 81–112.

Hobbes, Thomas. *Leviathan*. Ancient Wisdom Publication, 2019.

Knuth, Donald Ervin. *The Art of Computer Programming*. Addison-Wesley, 2010.

Lautman, Albert, and Simon B. Duffy. "Mathematics, ideas and the physical real." *Continuum*, 2011, 197-220.

Lévi-Strauss Claude. *Tristes Tropiques*. Translated by John Weightman and Doreen Weightman. Penguin Books, 2012.

Mauro W. Barbosa de Almeida, et al. "Symmetry and entropy: Mathematical metaphors in the work of Levi-Strauss [and Comments and Reply]." *Current Anthropology*, vol. 31, no. 4, 1990, pp. 367–385. *JSTOR*, www.jstor.org/stable /2743257. Accessed 2 May 2020.

Morozov, Evgeny. *The Net Delusion: How Not to Liberate the World*. Penguin, 2012.

Prigogine, Ilya. "The networked society." *Journal of World-Systems Research*, vol 5, no 3, 2000, pp. 892–898.

Slack, Jennifer Daryl, and J. Macgregor Wise. *Culture Technology: a Primer*. Peter Lang, 2007.

Smith, Adam. *Wealth of Nations*. Vintage, 2020.

Turing, A. M. "Computing machinery and intelligence." *Mind*, vol. 59, 1950, pp. 433–460.

Valiant, Leslie G. "Evolvability." *Journal of the ACM (JACM)*, vol. 56, no. 1, 2009, pp. 1–21.

Vazirani V. V. "Can the theory of algorithms ratify the "Invisible Hand of the Market"? In: Hirsch E. A., Karhumäki J., Lepistö A., Prilutskii M. (eds), *Computer Science – Theory and Applications*. CSR, 2012. Lecture Notes in Computer Science, vol. 7353. Springer, Berlin, Heidelberg.

Wardrip-Fruin, Noah, and Nick Montfort. *The NewMediaReader*. MIT Press, 2010.

Wiener, Norbert. "Men, machines, and the world about them," *WNYC*, October 25, 1950. https://www.wnyc.org/story/men-machines-and-the-world-about-them/

Chapter 5

Machines of Liberation, Machines of Control

The Ambiguous Roots of Data Capitalism

Reka Patricia Gal

THE HACKING OF THE SOCIAL

Whether we think about heist movies featuring a hacker to infiltrate systems, or TV shows and motion pictures focused on computing, the figure of the hacker in contemporary popular culture often emerges as an ethically ambiguous security professional able to infiltrate computer systems and use their technological savvy in security-related activities—for either benevolent or nefarious ends. This new image of the hacker is vastly different from the original, first-generation figures of the 1960s computer labs of the Massachusetts Institute of Technology (MIT), where they were imagined to be the subversive antiauthoritarian programmers who used computers for the betterment of humanity. The aim of hacking, as articulated in this early stage, was to avoid corruption. The hacker ethic was an ethic based on ideas of openness and technological innovation by institutional forces.[1] In the current digital economy, the "real life" hackers in Silicon Valley deploy the hacker ethic to "move fast and break things,"[2] capitalizing on creativity in order to disrupt markets, transforming the hacker ethic's innovative potential into serving market dynamics. In a sense, hacking became about quick market profits and proprietary software. The hacker ethic, concerned with finding technological solutions to everyday problems, has proven useful for finding and creating new market opportunities.

As Steven Levy shows in his book *Hackers: Heroes of the Computer Revolution*, which was first published in 1984, this transition occurred as the original MIT hackers were recruited by Stanford University's *Artificial Intelligence Lab* and as the subsequent generations of hackers started working for tech companies or started their own businesses. Among these were Stewart Nelson and Mike Lewitt who founded Systems Concepts, which was responsible for the movement of a large number of hackers from MIT to San Francisco.[3] The absorption of these hackers by the culturally bohemian Bay Area and the simultaneous rise of neoliberalism as a political and economic ideology greatly contributed to the development of Silicon Valley as the leading region of technological innovation and venture capital. It also provided a fertile ground for the techno-utopian culture Richard Barbrook and Andy Cameron have termed the "California Ideology." This techno-utopia functions as "a contradictory mix of technological determinism and libertarian individualism" that constitutes "the hybrid orthodoxy of the information age."[4]

New ubiquitous technologies—social media, *Google*'s many applications, *Roombas*, *Fitbits*, and so on—are not only providing comfortable fixes and shortcuts for daily activities but are also collecting and economizing all of their informational in- and output. *Roombas* memorize floor plans[5] and *Google* saves all search queries.[6] An array of information technology devices around the world are responsible for making it possible to aggregate and navigate through all this data and further transform the Information Age into the Big Data Age. Although the term "big data" is used often, a universally accepted definition has yet to develop. Broadly defined, big data refers to a specific composite of technology able to process vast quantities of data and the mining of these data sets for patterns as well as "distilling the patterns into predictive analytics and applying the analytics to new data."[7] Social psychologist Shoshana Zuboff claims that big data is "the foundational component in a deeply intentional and highly consequential new logic of accumulation," which "aims to predict and modify human behavior as a means to produce revenue and market control."[8] Zuboff posits that this new logic of accumulation has led to the restructuring of capitalism into what could be called *surveillance capitalism.*[9] The techno-utopian hacker ethic has helped make computational devices ubiquitous and hence contributed to the scaffolding of this new mode of accumulation. Yet, in surveillance capitalism, the original hacker ideal of creating technologies as fronts of human liberation has been turned around, turning them into machines of control instead.

Other scholars have also argued that the currently dominating principles of the digital economy amount to a new capitalist mode of production, which has been given various names: while Zuboff terms it surveillance capitalism,[10] McKenzie Wark refers to it as information capitalism,[11] and Nick Srnicek christens it platform capitalism.[12] Scholars such as Pekka Himanen[13] and Lily Irani[14] have shown how hacking and entrepreneurialism are intimately connected, while scholars such as Ruha Benjamin,[15] Sarah Sharma,[16] and Safiya Noble[17] have argued that technologies are explicitly setting the parameters for the social. Building on the work of these scholars, I posit that the absorption of the hacker ethic into the modus operandi of Silicon Valley neoliberalism is directly responsible for the emergence of a new type of high-tech company around the globe that uses data mining tactics in a way that can be considered to be going further than the hacking of technological systems and thus belongs in a different realm: that of the *hacking of the social*. These practices turn the original utopian hacker ideal of creating machines as fronts for human liberation into a new manifesto for political and social control through algorithmic guidance. In outlining the transition from hacking for liberation and toward hacking for social control, I center the "turn to the dark side" of the hacker, the programmer, and the engineer into data capitalist or data entrepreneur as a microcosm of the historical transition to neoliberalism. Focusing on the figure of the hacker will allow me to highlight how the hacker ethic has been turned against itself in surveillance capitalism, while seeming to realize its highest principle: "Information should be free." I guide the reader through two historical moments in which the hacker ethic becomes harnessed for surveillance capitalism. First, I will explore how hacking principles were instrumentalized in capital's attempts to restart accumulation after the 1970s crisis on new technological grounds, eventually resulting in the "dot com bubble." Second, I will analyze the turn to capturing and profiting from social behavior in today's Big Data Age.

CALIFORNIA DREAMING

The contemporary image of the hacker is heavily influenced by popular cultural depictions that showcase hackers as computer experts who induce security breaches and aid heists.[18] When it comes to depictions of hackers not based in fiction, the hacking group *Anonymous* is probably most well known, followed by some of the twenty-first-century whistleblowers Edward Snowden and Chelsea Manning whose activities are connected to bringing to light national and global surveillance programs.[19] Also of note is the subject of the 2018 *Cambridge Analytica* data scandal, Christopher Wylie.

This rooting of the figure of the hacker in security-related activities fundamentally changes and potentially misplaces the original meaning of the hacker and of the hacking activity. The term "to hack" originates from what can be considered the "birthplace" of hackers: MIT's Tech Model Railroad Club (TMRC) founded in 1946-founded members were among the first computer programmers. The members of TMRC used the phrase to describe "a project undertaken or a product built not solely to fulfill some constructive goal, but with some wild pleasure taken in mere involvement."[20] As the members of MIT's TMRC, and later its Artificial Intelligence Lab, were passionate about—some might say even obsessed with[21]—computers and finding ever newer possibilities in computing, they came to refer to themselves as hackers.

In its origins, hacking was conceived of as an apolitical activity centering around creativity and *innovation*. McKenzie Wark goes so far as to refer to any type of innovation, any type of creative manipulation of a system's rules, as hacking (although she does point out that the term originates in electrical engineering and computing). For Wark, hacking entails "whatever code we hack, be it programming language, poetic language, math or music, curves or colorings, we are the abstracters of new worlds."[22] Hackers bring about a new version of the system; they play with its rules and manipulate them until an entirely new "world" emerges. This focus on creativity and innovation is still evident in the usage of the term "hack" in some contemporary examples, such as, the term "life hack," which is used to define creative solutions that increase efficiency in different spheres of life and in the term "Hackathon," which is a programming event that encourages computer programmers and software developers to collaborate on solving a technological problem through creative, unusual solutions.[23]

The *Hacker Ethic*, or what it takes to be recognized as a hacker, is not something the hacker community officially agreed and put into a manifesto. As Levy described it, these rules were rather agreed silently and had evolved organically during the activities of these programmers. He gathered their main ethical rules as follows:

1. Access to computers—and anything which might teach you something about the way the world works—should be unlimited and total. Always yield to the Hands-on Imperative!
2. All information should be free.
3. Mistrust authority—promote decentralization.
4. Hackers should be judged by their hacking, not bogus criteria such as degrees, age, race, and position.
5. You can create art and beauty on a computer.
6. Computers can change your life for the better.[24]

What can easily be discerned from this account is exactly that type of techno-utopian vision that Richard Barbrook and Andy Cameron claim constitutes the contemporary foundation of the digital economy. This vision, however, was founded in liberal and antiauthoritarian ideals. For these first-generation hackers, free access and freedom of information could have constituted an ideal future of emancipation via technology where everyone would have had the possibility to hack, to create something new with computers and find joy in it. One of these first-generation hackers, Richard Stallman calls the antiauthoritarian nature of this movement a "constructive anarchism," which he says, however, "does not mean advocating a dog-eat-dog jungle. American society is already a dog-eat-dog jungle, and its rules maintain it that way. We [hackers] wish to replace those rules with a concern for constructive cooperation."[25] Fueled by anti-war sentiment,[26] the first-generation hackers originally conceived of their movement as a revolution of human and machine cooperation, directed against individualism and authoritarianism.

Creation and cooperation: the hope was that these machines would serve as vehicles of liberation for humans. The utopian ideal already started to crumble in the 1960s, as the hacker ethic spread, and some of the hackers from MIT's AI Lab moved to Stanford University. The corruption of hacker idealism was almost inevitable, claims Steven Levy: "As the computer revolution grew in a dizzying upward spiral of silicon, money, hype, and idealism, the Hacker Ethic became perhaps less pure, an inevitable result of its conflict with the values of the outside world."[27] While Levy features a euphemistic "perhaps" it is arguable that it was exactly this fusion of technology, money, and hype that resulted from the spread and corruption of the hacker ethic and led to what Silicon Valley is today and, consequently, to the type of surveillance capitalism of the current digital economy. In fact, in a second 1994 edition of his book Levy revisits this topic and claims in the afterword: "I also saw the emergence of a new wave: the present-day heirs to the hacker legacy who grew up in a world where commerce and hacker were never seen as opposing values. They are molding the future of the movement."[28]

From Levy's perspective, the original utopian image of the hacker as an antiauthoritarian liberator of humans has been co-opted by Silicon Valley capitalism. But hacking is rooted in innovation, and innovation implies that hacking ushers in a new way of designing *economic* value. The hacker as a figure whose modus operandi, if generalized, is one of "abstracting of new worlds," is extremely similar to economic Joseph Schumpeter's definition of the *entrepreneur*. Schumpeter writes: "The entrepreneur and his function are not difficult to conceptualize: the defining characteristic is simply the doing of new things or the doing of things that are already being done in a new way (innovation)."[29] In fact, Pekka Himanen connects the hackers' unwavering love of working on computing problems to Max Weber's *The Protestant Ethic and the Spirit of Capitalism,*[30] while Lily Irani argues that contemporary hackathons fulfill "a pedagogy of entrepreneurialism to manage the politics and energies of . . . development."[31]

But the "corruption" of the hacker ethic, regardless of how antiauthoritarian it was originally perceived to be, is also easier to comprehend once we take a closer look at the institutional architecture that made computational innovations possible. MIT was the largest nonindustrial defense contractor in the United States after World War II,[32] with most funding flowing into research on electronics, radars, and nuclear technology, which has undoubtedly made the intellectual as well as organization structures of science and engineering education fit the needs of national security.[33] As such, despite the anti-war and anarchistic aspirations of the original hackers, both the research in the 1946 formed TMRC, as well as its 1970s iteration, as the Artificial Intelligence Laboratory, have been made possible through U.S. Department of Defense funding. Based in military funding,[34] mediated through university science, mathematics, and engineering departments, and existing in a global economy ruled by capitalist logic, hacking was perhaps predestined to be co-opted by market dynamics.

The continuous emergence of new networked technologies—from PCs and the Internet through mobile phones and smart watches, and so on—have led to new organizational forms and markets, and contributed to reshaping the capitalist economic system as well. As Zuboff, Srnicek, and Wark have pointed out, the development of information and communication technologies (ICTs) has led to a new mode of capitalist production—and concurrently, a new type of capitalism.

In the contemporary moment, where the startups turned technology giants *Google*, *Facebook*, *Apple*, *Amazon*, and so on[35] are leaders in the digital economy, Levy's forecast of the merging of the market and hacker logics seems particularly relevant. In order to understand how Silicon Valley became the center of technological innovation that it is today, not only do we have to understand how the hacker ethic emerged from the culture of the 1960s, but we also have to consider how the force of neoliberalism simultaneously rose with it. Neoliberalism, as a political and economic concept, is a highly contested, "loose and shifting signifier"[36] due to its variance in space and time. Here, I use Wendy Brown's definition of neoliberalism to refer to a "form of reason that configures all aspects of existence in economic terms"[37] and the "discursive formulations, policy entailments, and material practices"[38] supporting it. Neoliberalism aims to privatize and economize noneconomic realms, to expand the logic of capitalist markets into everyday life. As hackers aimed to expand the reach of computers and make them freely available for everyone, neoliberalist structures recognized this potential and latched on to these technological developments so as to penetrate households. Today, we are living in a time of ubiquitous computing in which computers pervade virtually every sphere of life: not only is it the desktop computer one might have in the average living room or the laptop we can carry everywhere. Our smartphones, fridges, hoovers, and even our glasses are computing. The data generated through these devices makes not only big data architectures possible but also the transformation of hacking from its originally conceived orientation toward purely computational media, and toward the social itself.

Simultaneously, these devices, as well as the conception of new forms of these devices, are subsumed in the market structures and, hence, in the economic system of capitalism, which aims to extract surplus value for the capitalist out of the labor of the worker. Nick Srnicek traces the currently prevailing logic of capitalist accumulation—that is, the rise of platforms that aim to capitalize on big data—to three major occurrences in recent history, all of which are deeply connected to technological developments: the 1970s downturn; the dot-com boom and bust of the 1990s; and the reaction to the 2008 financial crisis. The 1970s financial crisis, also referred to as the Oil Crisis Recession, occurred due to the market dependence of capitalist organizations. While workers in pre-capitalist formations had direct access to their means of production, "under capitalism . . . economic agents are now separated from the means of subsistence and, in order to secure the goods they need for survival, they must now turn to the market."[39] He further posits that as production became market-oriented, businesses resorted to reducing production costs relative to prices, which could largely be achieved through adopting technologies that would drive the effectiveness of production. The 1990s so-called "dot-com boom and bust," also referred to as the "dot-com bubble," signified a time of economic frenzy regarding the increased usage and possibilities of networked computers, which effectively led to heightened investment and commercialization of a sphere of life which has previously been mainly noncommercial, namely the Internet. And at last, the 2008 financial crisis, which developed from a mortgage crisis in the United States, became a global banking crisis. Central banks reacted to the following recession by creating low-interest-rate environments, which in turn pushed investors toward more and more uncertain assets—such as developing unproven tech companies.[40]

This economic environment became conducive to the emergence of the currently prevailing global startup culture, ruled by what can be summarized in *Facebook* CEO Mark Zuckerberg's quote, "Move fast and break things. Unless you are breaking stuff, you aren't moving fast enough,"[41] which simultaneously unites the techno-utopian solutionist idealism of the hacker with the needs and wants of capitalism: creation, expansion, and gain. Melded together, they produced The Californian Ideology, which "simultaneously reflects the disciplines of market economics and the freedoms of hippie artisanship."[42]

This new logic of innovation under the guise of hacking brings together two opposites, or at least values that were meant to be opposites. McKenzie Wark has criticized this fusion in her *A Hacker Manifesto* (2004): "It is in the nature of hacking to discover freely, to invent freely, to create and produce freely. But it is not in the nature of hacking itself to exploit the abstractions thus produced."[43] The absorption of the originally antiauthoritarian and anti-market Hacker Ethic by capitalist organizational forms demonstrates exactly the success of neoliberalism in economizing all spheres. It manages to penetrate and economize even those spheres and producers whose values should be explicitly opposed to it.

Wendy Liu also laments on the deeply embedded capitalism of Silicon Valley in her memoir of her time as a *Google* intern and argues that while the company capitalizes on the *ethos* of hacking, workers are explicitly discouraged from working on open-source software.[44] First-generation hackers Richard Greenblatt and Richard Stallman also actively criticize the change from the original Hacker Ethics to the currently prevailing ones.[45] Not only has the commercialization of the Internet led to copyright notices that oppose the idea of the free flow of information and openness that the Hacker Ethic relies on, but it has also found ever newer ways to exploit Internet usage. Stephen Levy quotes Greenblatt on this:

> The real problem, Greenblatt says, is that business interests have intruded on a culture that was built on the ideals of openness and creativity. . . . There's a dynamic now that says, "Let's format our web page so people have to push the button a lot so that they'll see lots of ads," Greenblatt says. "Basically, the people who win are the people who manage to make things the most inconvenient for you."[46]

Of course, inconvenience is no longer the issue.[47] Greenblatt laments the way Internet-based companies rely on a business model that is based on advertising toward a user base. Advertising revenue has been a secure source of income for these companies. As Nick Srnicek argues, in the wake of the dot-com crisis, these businesses turned to monetize free resources, namely the data generated by their user base.[48] Twenty-first-century capitalism has turned data into a commodity and social life into a technology to be hacked.

The turn from disciplinary societies toward societies of control, combined with the turn from treating members of the society as individuals toward treating them as data points, or "dividuals," was already argued by Gilles Deleuze in 1992.[49] The unique combination of liberty and control that unfolds under contemporary data capitalism is one that, as Wendy Chun and Byung-Chul Han have pointed out in their respective works, utilizes the guise of freedom on the Internet to manage populations.[50] While Chun still refers to this type of control as *biopolitics*, Han posits that this type of control should be understood as *psychopolitics*: "Today, we are entering the age of digital psychopolitics. It means passing from passive surveillance to active steering. As such, it is precipitating a further crisis of freedom: now, free will itself is at stake."[51] Under data capitalism, control is baked into the foundations of technologies, making it invisible and therefore extraordinarily persuasive.

This makes it possible to guise this new type of control *as* freedom. It is, after all, the technology user that makes the choice in the end.

This type of surveillance has had especially harmful outcomes for Black and other racialized communities. Data aggregation under surveillance capitalism continues to afflict disproportionate harm on communities of color, as data-driven justice mechanisms lend the voice of perceived objectivity to predictive policing decisions. Information studies scholar Safiya Umoja Noble's *Algorithms of Oppression* (2018) shows that *Google* and *Facebook*'s algorithmic profiling, which is carried out based on data aggregated by users on their websites, shows both racist and gendered bias,[52] while sociologist Ruha Benjamin has shown that even the technologies that aim to critique and highlight racial hierarchies often end up explicitly intensifying these instead.[53] That is to say, even the technologies that are developed with the hope of having liberatory, equity-serving potential, end up serving the *status quo*.

Legal scholar Jonathan Zittrain echoes Chun and Han's criticisms in his summary of the problems with the present trajectory of Internet-based technological developments:

> The PC revolution was launched with PCs that invited innovation by others. So too with the Internet. . . . But the future unfolding right now is very different from this past. The future is not one of generative PCs attached to a generative network. It is instead one of sterile appliances tethered to a network of control.[54]

Drawing on her analysis of ex-*Google* employee James Damore's internal memo which advocated for tech as an inherently male space, and the recent rise of Mommy-apps aimed at entrepreneurializing care work, media theorist Sarah Sharma highlights that contemporary programmers in Silicon Valley are aware that their innovations are explicitly bound up in power structures and setting the parameters for the social.[55] While the first-generation Hackers dreamed of a world of creation and freedom ushered in through apolitical hacking activities, twenty-first-century hacking has been co-opted by corporate interests where control *is* the point. Openness has been exchanged for closed systems as the technological innovations of hackers started serving, first, the interest of national security, and, then, of capital. Or as Steven Levy, quoting Richard Stallman about the absorption of the hacker ethic into Silicon Valley neoliberalism, says, "They stole our world, . . . and it's irretrievably gone."[56]

NETWORK OF CONTROL

"Freedom with and of information is the utopia of the hacker class,"[57] states Wark in her *A Hacker Manifesto*, and this stands true even for the current generation of hackers. But the word "freedom" has become a shifting signifier. There are, of course, still hackers out there fighting for the freedom of information in the form of, for example, open-source software, but for the majority of computer programmers who are integrated into the workings of Silicon Valley, freedom of information has been transformed into freedom *to* information. Here I am not only referring to the extraction of data but also to the way these companies go about asking for—or rather *not* asking for—permission for data gathering processes (as is made evident by the hundreds of cases launched against *Google* regarding issues of privacy).[58] As Zuboff argues, this type of data gathering is evident in, for example, the business model of *Google* Street View: a *Google* car drives around and records images of the street, but the company has not asked permission to photograph millions of streets and homes as well as gather Wi-Fi information from on the street. "It simply takes what it wants."[59]

This contortion from "freedom of" toward "freedom to" is what stands at the center of the smooth slide of the figure of the hacker into that of the data entrepreneur. The Hacker Ethic, which demands that information be free so that cooperative innovation can occur around the world, is corrupted by the very fact that this innovation takes place under the auspices of capitalism. And once more, the similarities between the figure of the hacker and the figure of the entrepreneur are highlighted. As Isabelle Stengers writes, the entrepreneur "*demands the freedom to be able to transform everything into an opportunity*—for new profits, including what calls the common future into question."[60]

Once a startup embracing the logic of hacking, *Google* has become one of the biggest companies dealing in big data: it gathers data from its users, abstracts, aggregates, analyzes, and sells it. There is no reciprocity with the user: this "is a one-way process, not a relationship."[61] As Karen Yeung and Zuboff have pointed out in their respective works, this is further complicated by the lack of dialogue and consent from the side of the users, with Yeung arguing that, even in the case of online privacy notices, "individuals are highly unlikely to give meaningful, voluntary consent to the data sharing and processing activities entailed by big data analytic techniques," because they neither read nor understand these, because of the aggregation of businesses that deal in data gathering, and because privacy preferences might also be context-dependent.[62]

Information is free, in that it is given away freely, and free to use by these companies—as well as by the hackers of these companies, out of which to create new solutions. As Stengers put it, "Capitalism knows how to profit from every opportunity."[63] However, the information gathered from this aggregation of data is utilized not only for the creation of new markets. The information extracted from big data contains considerable insight into human behavior, which has made it possible to calculate, and predict human actions, and, as the emergence of companies such as *Cambridge Analytica* shows, even promises the possibility to *control* human actions. The techno-utopian promise of the California Ideology has become solving the "problem" of human behavior, the creativity and innovations of hackers harvested to further neoliberalist interests: configuring the citizens as consumers, and managing populations to further business—or other political—interests.[64]

In the age of Big Data, the act of hacking still centers around technological creativity: understanding how a system works and figuring out new and innovative workarounds through the manipulation of its rules. With the algorithmic decision guidance techniques made possible under surveillance capitalism, the rules of the system that are learned and manipulated are not only that of technology but also that of human behavior. Similarly to Zuboff, Yeung describes big data as "a mode of 'design-based' regulation,"[65] and Han posits that under big data "persons are being positivized into things, which can be quantified, measured and steered."[66] The hack has become an explicit hacking of the *social*, turning extracted information into an instrument of control. As Sharma points out in "Going to Work in Mommy's Basement," programmers are always tinkering with the social, as their understanding of the world always gets programmed into the technologies they invent.[67] And with algorithmic decision guidance techniques, this tinkering with the social becomes explicit.

Under a surveillance capitalist system, one could say that the hacker as a security professional is doing "behavioral security" work. This type of control stands in contrast to the extrinsic and spatialized disciplinary practices philosopher Michel Foucault refers to in his analyses of biopower.[68] Following the postulations of Han, neoliberal control is intrinsic and non-spatialized, the type that, as Yeung says, can be engineered through utilizing the insights into human behavior gained from big data and nudging the user into the wanted direction. She explains that big data driven guidance technologies are "self-contained cybernetic systems," operating through the cultivation of the user's "choice environment," meaning what they see as possible options online, giving feedback to the "choice architect" and continuous overseeing and cultivation of the "choice environment . . . in light of population-wide trends."[69] Algorithmic decision guidance is real-time, dynamic, and, personalized, making it virtually imperceptible and therefore extremely powerful.

Users are being fed "suitably adapted" choice possibilities online, carefully nudging them toward the consumption of goods, and this type of control is intensified under surveillance capitalism. The continuous recording and monitoring of user activities allows companies and institutions to create "psychological profiles" of users and predict, as well as alter, their behavior. Research conducted by Kosinksi et al. shows that "easily accessible digital records of behavior, *Facebook* Likes, can be used to automatically and accurately predict a range of highly sensitive personal attributes,"[70] while research by Matz et al. shows that using these for "psychologically tailored advertising . . . significantly altered [user] behaviour as measured by clicks and purchases."[71] Kosinski et al. note that these findings "may have considerable negative implications, because [they] can easily be applied to large numbers of people without obtaining their individual consent and without them noticing."[72]

No other contemporary event has made this clearer than the 2018 *Cambridge Analytica* scandal, where a political consulting company may have acquired approximately 87 million *Facebook* users' data and used them for psychological profiling in order to influence voter behavior during the Brexit campaign and the United States elections of 2016.[73] The research used as a basis for these practices is exactly that of Kosinski et al. and Matz et al.[74] The deployment of a *Facebook* application that received access to user's profiles and likes made it possible for the company to psychologically target them. If hacking is about understanding the rules of a system and learning ways to manipulate it via technological innovation, then *Cambridge Analytica*'s big data-enabled, psychologically tailored algorithmic decision guidance techniques it has become (in)famous for are exactly that: *the hacking of the social.* The logic of hacking has not only been distorted by its absorption into neoliberalism, but it has also turned global in this process.

The hacker *qua* whistleblower of the story is Christopher Wylie, the software engineer who conceived the idea for Cambridge Analytica and who oversaw its operations. Wylie was always interested in politics. While still a student, he created an application for *Facebook* that determined users' personality traits based on *Facebook* likes, which went viral. Years later, when he read the research by Kosinski, he claims that something "clicked" for him. He realized how this could be used for political campaigns. He presented an idea for psychological targeting online to the Liberal Party of Canada, which was not interested,[75] but a connection there introduced him to the *SCL Group*, which is the company that went on to create *Cambridge Analytica*—based on the research by Kosinski et al. and with Wylie's computer expertise.[76]

For all intents and purposes, what Wylie created here is a hack: a technological innovation based on learning the rules of a system and manipulating them, with the rules here being both technological *and* psychological. As for the Hacker Ethic, the first two implied rules named by Levy have made surveillance capitalism, and this type of psychological targeting, possible, even if in a twisted, morally questionable way: "1. Access to computers should be unlimited and total. 2. All information should be free."[77] It is precisely the heightened access to computing devices and the insistence of data companies to have the freedom *to* information that makes the aggregation of enough data possible that allows for insights into human behavior and the deployment of algorithmic decision guidance techniques to influence them on the very platforms with which they share their information. Wylie's "behavioural engineering" hack utilizes a new way of designing economic value, one which is beneficial to tech companies around the world who extract the data generated by their users. Wylie's hack is also politicizing the hacking activity, not only because it is implicated in political parties' election campaigns but also because it is *explicitly* aimed at meddling with power structures and the hacking of the social.

"Whatever code we hack, . . . we are abstractors of new worlds," says Wark in her manifesto. With this specific innovation, the combination of data mining and psychological targeting, Wylie abstracted a new world of algorithmic and psychological control where companies such as *Cambridge Analytica* can thrive. Subsequently, he became one in the line of hacker whistleblowers who are calling attention to the abundance and pervasiveness of surveillance tactics both by governments and by private companies. Except, unlike the cases of Snowden or Manning, the case of Wylie and *Cambridge Analytica* could be seen as the end result of the corruption of the hacker ethic by economic forces: an indication that the techno-utopian vision of the original hackers is virtually unobtainable under a neoliberal economic and political system.

SUBVERSIVE FUTURES

What options, then, are still left if we wish to combat this neoliberalization of the hacker ethic and develop technologies with liberatory power? For one, in combatting surveillance capitalism and protecting privacy and security on the Internet, cryptography could be a powerful practice. Knowing how to analyze and construct Internet protocols so only the sender and receiver, and no third party, can access it. And, in light of the *Cambridge Analytica* scandal, simply knowing how to use the Internet in a way that does not allow for the extraction of data from everyday daily use are convenient means to combat surveillance capitalist structures. There are a number of organizations that wish to share their knowledge in the area, mainly in the form of so-called crypto parties, organized, for example, by *CryptoParty*, which is a "decentralized, global initiative to introduce basic tools for protecting privacy, anonymity and overall security on the Internet to the general public." The organization makes clear that they "are neither commercially nor politically aligned," and that no NGOs, companies, or political parties may sponsor the organization,[78] which is an effective way of bypassing outside influence and the neoliberalization of the group.

A way to sidestep economization and campaign for freedom of access and freedom of information is creating and advocating for open-source software:

computer software the source code of which is made openly available only for studying, modification, and further noncommercial distribution.[79] The intellectual foundation of the open-source movement in fact originates from the first-generation hackers, most notably with Richard Stallman's *Free Software Foundation* founded in 1985.[80] Open-source software both directly works against the commercialization of hacking activities and aims to foster cooperation and decentralized hacking by making software easily accessible and modifiable by users. With that said, this would also have to be done in a way that prevents open innovation practices—such as open-source software and open data—from providing disproportionate value to existing powerful institutions by providing them with the free technological infrastructure to build with their enormous pools of preexisting resources and labor. One way to counter this would be to follow Wendy Liu's suggestion of reclaiming entrepreneurship from Silicon Valley capitalism by instituting state policies that transform technological innovation to serve the public.[81]

At last, we can say that whistleblowing is a useful practice, not within hacking itself, but in order to bring public awareness to underlying socio-economic tendencies and to promote social change. The whistleblowing of Edward Snowden, Chelsea Manning, and more recently of Christopher Wylie ushered in articles of introspection, calling attention to the power governments and companies qua technological innovation exercise over society. Wylie's whistleblowing, which started the Cambridge Analytica scandal, now explicitly brings attention not only to questions about data privacy but also to algorithmic decision guidance techniques, to societal *control*. What is questionable, however, is to what extent subsequent activist work has been effective at converting attention into action.

In all likelihood, in addition to the combination of these practices, fundamental institutional and economic reforms may be needed if we are to combat surveillance capitalism on a societal scale. Cryptography, open source software, and whistleblowing are important and helpful, but none of these will alone turn around a new global economic system and the tendency to extract data and use the data to algorithmically guide users. The issue will eventually have to be handled on the policy level, but in the meantime, these practices could strengthen the societal support for such policies as well as create a foundation for a more open and secure Internet. One thing is clear; however, steps toward a solution are urgently needed.

The extended logic of data capitalism can be traced to the origins of computing itself, to the first-generation programmers at MIT's TMRC who eventually moved to Stanford's Artificial Intelligence Lab in the 1960s. These programmers, or "hackers" as they called themselves, believed in the liberatory power of computers as part of their unspoken principles, the so-called Hacker Ethic. While hacking was originally conceived of as antiauthoritarian and anti-capitalist, its concern with finding technological solutions to everyday problems has proven useful for finding and creating new market opportunities, which has led to its being co-opted into neoliberal market structures. The convergence of the libertarian historical origins of their hacker ethic and the dissident culture of the early Internet has allowed for Silicon Valley's original focus on creativity and emancipation to be turned toward associations we are now accustomed to make with libertarianism: extreme individualism and market fundamentalism. By the late 1990s, the hacker ethic was being instrumentalized in capital's attempts to restart accumulation after the 1970s crisis on new technological grounds, resulting in the "dot com bubble" and the 2008 Financial crisis, which further pushed investors toward investing in tech startups. I argue that the hacker ethic is also being deployed in the surveillance capitalist turn toward capturing and profiting from social behavior today. The algorithmic decision guidance techniques accompanying current big data architectures have only become possible due to a scaffolding of ubiquitous computing, itself a result of the Hacker Ethic's and capital's conjoined push toward making access to computational devices total. Under surveillance capitalist algorithmic guidance structures, the original hacker ideal of creating technologies as fronts of human liberation is distorted, turning these into machines of control instead. The 2018 *Cambridge Analytica* scandal makes it clear that hacking is no longer only concerned with the hacking of computational systems. Instead, hacking has been re-politicized; it has moved on to learning and manipulating the rules of human behavioral systems. The age of surveillance capitalism is, then, the age of the hacking of the social.

NOTES

1. Steven Levy. *Hackers: Heroes of the Computer Revolution*, 1st edition. Sebastopol, CA: O'Reilly Media, 2010.
2. Henry Blodget. "Mark Zuckerberg on Innovation," *Business Insider*, 2009. https://www.businessinsider.com/mark-zuckerberg-innovation-2009-10.
3. Levy, *Hackers: Heroes of the Computer,* 133.
4. Richard Barbrook and Andy Cameron, "The Californian Ideology," *Science as Culture* 6, no. 1 (1996): 48. doi:10.1080/09505439609526455.
5. Alex Hern. "Roomba Maker May Share Maps of Users' Homes with Google, Amazon or Apple." *The Guardian*, 2017, sec. Technology. http://www.theguardian.com/technology/2017/jul/25/roomba-maker-could-share-maps-users-homes-google-amazon-apple-irobot-robot-vacuum.
6. Alix Langone. "Even If You Clear Your History, Google Has a Record of All of Your Search Activity—Here's How to Delete It." *Business Insider*, 2018. https://www.businessinsider.com/even-if-you-cleared-your-history-google-records-your-search-activity-2018-4.

7. Karen Yeung. "'Hypernudge': Big Data as a Mode of Regulation by Design," *Information, Communication and Society* 20, no. 1 (2017): 118–36. doi:10.1080/1369118X.2016.1186713.

8. Shoshana Zuboff. "Big Other: Surveillance Capitalism and the Prospects of an Information Civilization." *Journal of Information Technology* 30, no. 1 (2015): 75–89. doi:10.1057/jit.2015.5.

9. Although in this paper I will rely on Zuboff's analysis and term, other scholars have also pointed out the existence of this new system in similar ways. See for example Nick Srnicek, *Platform Capitalism* (John Wiley & Sons, 2016) and McKenzie Wark, "What If This Is Not Capitalism Any More, but Something Worse? NPS Plenary Lecture, APSA 2015, Philadelphia, PA," *New Political Science* 39, no. 1 (2017): 58–66. doi:10.1080/07393148.2017.1278846.

10. Zuboff, "Big Other."

11. Wark, "What If This Is Not Capitalism Any More, but Something Worse?"

12. Srnicek, *Platform Capitalism.*

13. Pekka Himanen. *The Hacker Ethic, and the Spirit of the Information Age*. New York: Random House, 2001.

14. Lilly Irani, "Hackathons and the Making of Entrepreneurial Citizenship," *Science, Technology, and Human Values* 40, no. 5 (2015): 799–824. doi:10.1177/0162243915578486.

15. Ruha Benjamin. *Race after Technology: Abolitionist Tools for the New Jim Code*. John Wiley & Sons, 2019.

16. Sarah Sharma. "Going to Work in Mommy's Basement." Text, *Boston Review*, 2018. http://bostonreview.net/gender-sexuality/sarah-sharma-going-work-mommys-basement.

17. Safiya Umoja Noble. *Algorithms of Oppression: How Search Engines Reinforce Racism*. New York: New York University Press, 2018.

18. See for example the television shows *Mr. Robot*, *Criminal Minds*, or *CSI: Cyber*.

19. Gabriella Coleman. *Hacker, Hoaxer, Whistleblower, Spy: The Many Faces of Anonymous*. Verso Books, 2014.

20. Levy, *Hackers: Heroes of the Computer*, 10.

21. Levy, *Hackers: Heroes of the Computer*, 437–479.

22. McKenzie Wark. *A Hacker Manifesto*. Harvard University Press, 2004.

23. Briscoe, G., and Catherine Mulligan. "Digital Innovation: The Hackathon Phenomenon." Creativeworks London Working Paper No. 6, London's Digital Economy, 2014. http://www.creativeworkslondon.org.uk/wp-content/uploads/2013/11/Digital-Innovation-The-Hackathon-Phenomenon1.pdf.

24. Levy, *Hackers: Heroes of the Computer*, 27–28.

25. Levy, *Hackers: Heroes of the Computer*, 438.

26. Levy, *Hackers: Heroes of the Computer*, 124.

27. Levy, *Hackers: Heroes of the Computer*, 451.

28. Levy, *Hackers: Heroes of the Computer*, 466.

29. Joseph A. Schumpeter. "The Creative Response in Economic History." *The Journal of Economic History* 7, no. 2 (1947): 149–159. doi:10.1017/S0022050700054279.

30. Himanen, *The Hacker Ethic, and the Spirit of the Information Age*.

31. Irani, "Hackathons and the Making of Entrepreneurial Citizenship."
32. Stuart W. Leslie. *The Cold War and American Science: The Military-Industrial-Academic Complex at MIT and Stanford.* Columbia University Press, 1993, 14.
33. Leslie, *The Cold War and American Science*, 43.
34. Levy, *Hackers: Heroes of the Computer*, 59, 125.
35. For example, the companies of venture capitalist Peter Thiel, founder of PayPal, whose big data analysis company Palantir should be a subject to further analysis. See Waldman, Peter, Lizette Chapman, and Jordan Robertson. "Palantir Knows Everything About You." *Bloomberg*, April 19, 2018. https://www.bloomberg.com/features/2018-palantir-peter-thiel/.

www.bloomberg.com/features/2018-palantir-peter-thiel/ Accessed on May 1, 2018.

36. Wendy Brown. *Undoing the Demos: Neoliberalism's Stealth Revolution.* MIT Press, 2015, 20.
37. Levy, *Hackers: Heroes of the Computer*, 17.
38. Levy, *Hackers: Heroes of the Computer*, 20.
39. Srnicek, *Platform Capitalism.*
40. Srnicek, *Platform Capitalism.*
41. Blodget, "Mark Zuckerberg on Innovation."
42. Barbrook and Cameron, "The Californian Ideology," 50.
43. Wark, *A Hacker Manifesto*, Section 075. Wark goes on to say that that the realization of each hack traps it in the form of property and hence provides the hack with an assigned a value—it commodifies it. In the Manifesto, she argues in the Marxist tradition for an understanding of hackers as a class within the capitalist system, and for their collaboration with the proletariat against a new type of ruling class that is the owner of information and the pathways to information—the vectoralist class. These vectoralists are the ones who own the means of production, who are responsible for monopolizing on the creations of hackers, and hence responsible for the corruption of the Hacker Ethic through neoliberalist interests.
44. Wendy Liu. *Abolish Silicon Valley: How to Liberate Technology from Capitalism.* Watkins Media Limited, 2020.
45. See Levy, *Hackers: Heroes of the Computer*, 437–479.
46. Levy, *Hackers: Heroes of the Computer*, 470.
47. Or rather, inconvenience was the issue of late 2000s advertising. However, the shift to identity-based marketing through the 2010s re-focused digital ads on creating more convenient and personalized user experiences, which was itself made possible through new Big Data and Artificial Intelligence technologies.
48. Srnicek, *Platform Capitalism.*
49. Gilles Deleuze. "Postscript on the Societies of Control," *October* 59 (1992): 3–7.
50. See Wendy Hui Kyong Chun, *Control and Freedom: Power and Paranoia in the Age of Fiber Optics* (MIT Press, 2008) and Byung-Chul Han, *Psychopolitics: Neoliberalism and New Technologies of Power* (Verso Books, 2017).
51. Han, *Psychopolitics,* 11.
52. Noble, *Algorithms of Oppression.*

53. Benjamin, *Race after Technology.*

54. Jonathan Zittrain. *The Future of the Internet—And How to Stop It.* Yale University Press, 2008.

55. Sharma, "Going to Work in Mommy's Basement."

56. Levy, *Hackers: Heroes of the Computer*, 471.

57. Wark, *A Hacker Manifesto.*

58. See for example Clare Duffy, "Google Agrees to Pay $13 Million in Street View Privacy Case," *CNN*, 2019. https://www.cnn.com/2019/07/22/tech/google-street-view-privacy-lawsuit-settlement/index.html as well as Juliana Gruenwald Henderson, "Google and YouTube Will Pay Record $170 Million for Alleged Violations of Children's Privacy Law," Federal Trade Commission, September 3, 2019, https://www.ftc.gov/news-events/press-releases/2019/09/google-youtube-will-pay-record-170-million-alleged-violations.

59. Zuboff, "Big Other," 78.

60. Isabelle Stengers. *In Catastrophic Times: Resisting the Coming Barbarism.* Open Humanities Press, 2015, 66.

61. Zuboff, "Big Other," 79.

62. Yeung, "Hypernudge," 125.

63. Stengers, *In Catastrophic Times*, 11.

64. This raises interesting questions regarding the subject-object relations of humans and machines, and brings the utopian premise of technologically induced freedom into that of cybernetic control. A theoretical inquiry of algorithmic decision guidance techniques through the lens of sociocybernetics should be subject for further analysis.

65. Yeung, "Hypernudge," 118.

66. Han, *Psychopolitics*, 12.

67. Sharma, "Going to Work in Mommy's Basement."

68. See for example Michel Foucault and François Ewald, *"Society Must Be Defended": Lectures at the Collège de France, 1975–1976* (St Martins Press, 2003), and Michel Foucault, *Discipline and Punish: The Birth of the Prison* (Knopf Doubleday Publishing Group, 2012).

69. Yeung, "Hypernudge," 122.

70. M. Kosinski, D. Stillwell, and T. Graepel. "Private Traits and Attributes Are Predictable from Digital Records of Human Behavior." *Proceedings of the National Academy of Sciences* 110, no. 15 (2013): 5802–5805. doi:10.1073/pnas.1218772110.

71. S. C. Matz, M. Kosinski, G. Nave, and D. J. Stillwell. "Psychological Targeting as an Effective Approach to Digital Mass Persuasion." *Proceedings of the National Academy of Sciences* 114, no. 48 (2017): 12714–12719. doi:10.1073/pnas.1710966114.

72. Kosinski, Stillwell, and Graepel, "Private Traits and Attributes Are Predictable," 5805.

73. Christopher Wylie. "Christopher Wylie: Why I Broke the Facebook Data Story – and What Should Happen Now." *The Guardian*, 2018, sec. UK news. https://www.theguardian.com/uk-news/2018/apr/07/christopher-wylie-why-i-broke-the-facebook-data-story-and-what-should-happen-now.

74. Edmund L. Andrews. "The Science behind Cambridge Analytica: Does Psychological Profiling Work?" Stanford Graduate School of Business, 2018. https://www.gsb.stanford.edu/insights/science-behind-cambridge-analytica-does-psychological-profiling-work.

75. Andy Blatchford. "Liberals Awarded $100,000 Contract to Man at Centre of Facebook Data Controversy," *CBC News*, 2018. https://www.cbc.ca/news/politics/christopher-wylie-facebook-liberals-canada-cambridge-analytica-1.4586046.

76. Carole Cadwalladr. "'I Made Steve Bannon's Psychological Warfare Tool': Meet the Data War Whistleblower." *The Guardian*, 2018, sec. News. https://www.theguardian.com/news/2018/mar/17/data-war-whistleblower-christopher-wylie-faceook-nix-bannon-trump.

77. Levy, *Hackers: Heroes of the Computer,* 472.

78. Cryptoparty. "Guiding Principles." Accessed April 17, 2020. https://www.cryptoparty.in/guiding_principles.

79. Andrew M. St. Laurent. *Understanding Open Source and Free Software Licensing*. O'Reilly Media, Inc., 2004.

80. Levy, *Hackers: Heroes of the Computer,* 472.

81. Wendy, *Abolish Silicon Valley*, chap. 11, Kindle.

BIBLIOGRAPHY

Andrews, Edmund L. "The Science behind Cambridge Analytica: Does Psychological Profiling Work?" Stanford Graduate School of Business, 2018. https://www.gsb.stanford.edu/insights/science-behind-cambridge-analytica-does-psychological-profiling-work.

Barbrook, Richard, and Andy Cameron. "The Californian Ideology." *Science as Culture* 6, no. 1 (1996): 44–72. doi:10.1080/09505439609526455.

Benjamin, Ruha. *Race after Technology: Abolitionist Tools for the New Jim Code.* John Wiley & Sons, 2019.

Blatchford, Andy. "Liberals Awarded $100,000 Contract to Man at Centre of Facebook Data Controversy." *CBC News*, 2018. https://www.cbc.ca/news/politics/christopher-wylie-facebook-liberals-canada-cambridge-analytica-1.4586046.

Blodget, Henry. "Mark Zuckerberg on Innovation." *Business Insider*, 2009. https://www.businessinsider.com/mark-zuckerberg-innovation-2009-10.

Briscoe, G., and Catherine Mulligan. "Digital Innovation: The Hackathon Phenomenon." *Creativeworks London Working Paper No. 6*, London's Digital Economy, 2014. http://www.creativeworkslondon.org.uk/wp-content/uploads/2013/11/Digital-Innovation-The-Hackathon-Phenomenon1.pdf.

Brown, Wendy. *Undoing the Demos: Neoliberalism's Stealth Revolution.* MIT Press, 2015.

Cadwalladr, Carole. "'I Made Steve Bannon's Psychological Warfare Tool': Meet the Data War Whistleblower." *The Guardian*, 2018, sec. News. https://www.the

guardian.com/news/2018/mar/17/data-war-whistleblower-christopher-wylie-face ook-nix-bannon-trump.

Chun, Wendy Hui Kyong. *Control and Freedom: Power and Paranoia in the Age of Fiber Optics*. MIT Press, 2008.

Coleman, Gabriella. *Hacker, Hoaxer, Whistleblower, Spy: The Many Faces of Anonymous*. Verso Books, 2014.

Cryptoparty. "Guiding Principles." Accessed April 17, 2020. https://www.cryptopa rty.in/guiding_principles.

Deleuze, Gilles. "Postscript on the Societies of Control." *October* 59 (1992): 3–7.

Delfanti, Alessandro. *Biohackers: The Politics of Open Science*. London: Pluto Press, 2013.

Duffy, Clare. "Google Agrees to Pay $13 Million in Street View Privacy Case." CNN, 2019. https://www.cnn.com/2019/07/22/tech/google-street-view-privacy-la wsuit-settlement/index.html.

Ferguson, Andrew Guthrie. "Black Data: Distortions of Race, Transparency, and Law." In *The Rise of Big Data Policing*, pp. 131–142. Surveillance, Race, and the Future of Law Enforcement. NYU Press, 2017.

Foucault, Michel. *Discipline and Punish: The Birth of the Prison*. Knopf Doubleday Publishing Group, 2012.

Foucault, Michel, and François Ewald. *"Society Must Be Defended": Lectures at the Collège de France, 1975–1976*. St Martins Press, 2003.

Han, Byung-Chul. *Psychopolitics: Neoliberalism and New Technologies of Power*. Verso Books, 2017.

Henderson, Juliana Gruenwald. "Google and YouTube Will Pay Record $170 Million for Alleged Violations of Children's Privacy Law." Federal Trade Commission, 2019. https://www.ftc.gov/news-events/press-releases/2019/09/google-youtube-will-pay-record-170-million-alleged-violations.

Hern, Alex. "Roomba Maker May Share Maps of Users' Homes with Google, Amazon or Apple." *The Guardian*, 2017, sec. Technology. http://www.theguardian .com/technology/2017/jul/25/roomba-maker-could-share-maps-users-homes-goo gle-amazon-apple-irobot-robot-vacuum.

Himanen, Pekka. *The Hacker Ethic, and the Spirit of the Information Age*. New York: Random House, 2001.

Irani, Lilly. "Hackathons and the Making of Entrepreneurial Citizenship." *Science, Technology, and Human Values* 40, no. 5 (2015): 799–824. doi: 10.1177/0162243915578486.

Kosinski, M., D. Stillwell, and T. Graepel. "Private Traits and Attributes Are Predictable from Digital Records of Human Behavior." *Proceedings of the National Academy of Sciences* 110, no. 15 (2013): 5802–5. doi:10.1073/pnas.1218772110.

Langone, Alix. "Even If You Clear Your History, Google Has a Record of All of Your Search Activity—Here's How to Delete It." *Business Insider*, 2018. https:// www.businessinsider.com/even-if-you-cleared-your-history-google-records-your-search-activity-2018-4.

Laurent, Andrew M. St. *Understanding Open Source and Free Software Licensing*. O'Reilly Media, Inc., 2004.

Leslie, Stuart W. *The Cold War and American Science: The Military-Industrial-Academic Complex at MIT and Stanford.* Columbia University Press, 1993.

Levy, Steven. *Hackers: Heroes of the Computer Revolution*, 1st edition. Sebastopol, CA: O'Reilly Media, 2010.

Liu, Wendy. *Abolish Silicon Valley: How to Liberate Technology from Capitalism.* Watkins Media Limited, 2020.

Matz, S. C., M. Kosinski, G. Nave, and D. J. Stillwell. "Psychological Targeting as an Effective Approach to Digital Mass Persuasion." *Proceedings of the National Academy of Sciences* 114, no. 48 (2017): 12714–12719. doi:10.1073/pnas.1710966114.

Noble, Safiya Umoja. *Algorithms of Oppression: How Search Engines Reinforce Racism.* New York: New York University Press, 2018.

Schumpeter, Joseph A. "The Creative Response in Economic History." *The Journal of Economic History* 7, no. 2 (1947): 149–159. doi:10.1017/S0022050700054279.

Sharma, Sarah. "Going to Work in Mommy's Basement." Text. *Boston Review*, 2018. http://bostonreview.net/gender-sexuality/sarah-sharma-going-work-mommys-basement.

Srnicek, Nick. *Platform Capitalism.* John Wiley & Sons, 2016.

Stengers, Isabelle. In *Catastrophic Times: Resisting the Coming Barbarism.* Open Humanities Press, 2015.

Waldman, Peter, Lizette Chapman, and Jordan Robertson. "Palantir Knows Everything About You." *Bloomberg*, 2018. https://www.bloomberg.com/features/2018-palantir-peter-thiel/.

Wark, McKenzie. *A Hacker Manifesto.* Harvard University Press, 2004.

———. "What If This Is Not Capitalism Any More, but Something Worse? NPS Plenary Lecture, APSA 2015, Philadelphia, PA." *New Political Science* 39, no. 1 (2017): 58–66. doi:10.1080/07393148.2017.1278846.

Wylie, Christopher. "Christopher Wylie: Why I Broke the Facebook Data Story—and What Should Happen Now." *The Guardian*, 2018, sec. UK news. https://www.theguardian.com/uk-news/2018/apr/07/christopher-wylie-why-i-broke-the-facebook-data-story-and-what-should-happen-now.

Yeung, Karen. "'Hypernudge': Big Data as a Mode of Regulation by Design." *Information, Communication and Society* 20, no. 1 (2017): 118–136. doi:10.1080/1369118X.2016.1186713.

Zittrain, Jonathan. *The Future of the Internet—And How to Stop It.* Yale University Press, 2008.

Zuboff, Shoshana. "Big Other: Surveillance Capitalism and the Prospects of an Information Civilization." *Journal of Information Technology* 30, no. 1 (2015): 75–89. doi:10.1057/jit.2015.5.

Chapter 6

The Autoimmunitary Violence of the Algorithms of Mourning

Stefka Hristova

It is 2020 and I am working on this book chapter in the middle of a pandemic that has claimed more than 220,000 lives here in the United States. The high death toll, caused by the spread of the SARS-CoV-2 virus, has been publicized as a cautionary tale and reason for people to practice social distancing and has brought to the surface the fragility of human life and our own mortality. It has also been challenged by right-wing media as a hoax or as a sacrifice that the American public is willing to make in order to keep the economy going. While profuse media coverage of death in times when the actual death toll is high is appropriate, I want to address the ways in which social media tended to deliver clusters of content associated with death and suicide prior to the COVID-19 crisis.

In comparing the ways in which *YouTube* popularized and capitalized on Amanda Todd's suicide video *My Story* in 2012 to the ways in which *Facebook* and *Netflix* promote and monetize content related to death in 2019, I argue that there has been a significant shift both in what can be exploited for profit in social media as well as in how trending material is clustered and presented. In the Web 2.0 era media was marked by hardwired playlists and explicit dependence on commercial sponsors. Web 3.0 ushered in an era of the age of algorithmic media in which the media currency involved the interchangeable flow of content and commerce. The stakes of this shift are significant. By considering how death and mourning operate in social media in a pre-COVID-19 environment, I argue that algorithms of mourning not only lack but actively erase and capitalize on the cessation of humanity as they deliver what Jacque Derrida has theorized as autoimmunitary violence. Their rational—seemingly impartial calculations—create a world in which an algorithm relentlessly insists on giving us more of what it assumes we like instead of the little we might actually need. The algorithms of mourning structure a dangerous "filter bubble" governed by despair and hopelessness.

MEDIATED DEATH

I would like to begin my discussion on the ways that algorithms have changed how social media have capitalized on the topic of death by tracing *what* is at the core of the processes of commercialization. In the age of Web 2.0, social media platforms such as *Facebook, Myspace,* text message platforms were harnessed as places youth could mourn and remember friends through continuous communication.[1] As Margaret Gibson has argued, these mediated practices of mourning were already embedded in a process of automation. For her,

> automatic mourning thus refers to the ways in which the dead via new automated message services can posthumously reanimate their presence in the lives of loved ones repositioning themselves not just as the remembered dead but the dead who, by a strange reversal, are remembering the living. . . . Automated mourning expresses the idea of the self-propelling logic of user-generated media as people respond not just to events themselves but responses to responses.[2]

The automation in Gibson's study refers to the automated or standardized ways bereavement culture finds expression on social media: sad icons, standard messages, *Facebook* likes. This process of automation, I argue, extends beyond the boundary of the individual online memorial and beyond the individual instance of mourning. Rather, bereavement culture has come to be seen as participatory culture that can in turn drive revenue on the basis precisely of participation.

In fact, keeping social media participation even in the aftermath of one's death is still the preferred outcome for companies such as *Facebook*. In the wake of the increased COVID-19 death rate, the company recently raised the profile of its post-mortem "memorialization" options for its users moving it to the main settings menu. Given the increased probability of death in the current crisis, *Facebook* has attempted to secure the life of its users' accounts even past the physical lives of its users. "Your Legacy Contact" section asks users to choose a guardian for their account so that they can "manage tribute posts," "respond to new friend requests," and "update your profile picture and cover photo." A second, less preferred option is listed below as "Delete Account after Death." Users are discouraged to select this course of action as it signals the cession of online activity and thus appears as a threat to *Facebook*'s ad revenue stream.

Mediated death had become a golden revenue opportunity on social media via the promotion of both literal products and affective content. As Nikolaus Lehner has argued, the data and social media activity of deceased users away continue to inform the digital economy through what he calls the "*algorithmic undeath*."[3] What is compelling here is Lehner's theorization of the ways "the data of a deceased person who had an *Amazon* account could still be used to algorithmically process recommendations for another user who is still alive and well."[4] This harnessing of data is part of the everyday operations of digital capitalism, which "thrives more or less by secretly capturing and processing data traces people leave behind during their lifetime."[5] In thinking about processes of automation and algorithmically defined digital capitalism, I trace the violent process of the commercialization of mourning.

READ BLEACH, OFFER DETERGENT

On October 11, 2012, after being cyber-bullied for years by a stalker, Amanda Todd committed suicide. In the wake of her death, her *YouTube* video "My Story: struggling, bullying, suicide, self-harm" became an instant Internet hit and made her famous. Shot in a black and white aesthetic with Amanda appearing in black sandwiched between a white wall and the white note cards, the story evoked codes of authenticity and intimacy. Six minutes and thirty-six seconds into the video (6:36), we see a notecard containing the word "*bleach*." The card reads as follows:

I wanted to die so bad.
When he brought me home I drank bleach. . .

The consumer-driven algorithm picked up *bleach* as the relevant term and clustered its ads in terms of relevance to this consumer item. It therefore coupled the video with a banner ad for detergents centered and superimposed on Todd's video. The ad read "free laundry coupons" and featured seven different laundry detergents. This particular ad also appeared in a repost of Todd's video by the *YouTube* user ChiaVidoes as well as in subsequent reviews of the video. The blatant attempt of *YouTube* to capitalize on the popularity for Todd's video and the participatory online community became a hot topic of discussion on both *Reddit* and in the Internet meme community more broadly. As the *Reddit* thread "Too soon YouTube?" shows, the detergent ad placement was quite controversial. It generated 143 comments, most of which indicated irritation with the lack of sensitivity on the part of *YouTube* and suggested the use of ad blocking software as a solution. *YouTube*'s algorithm was attempting to capitalize on ad sales driven by increased user activity in a moment when a community was mourning. In the face of mourning, the algorithm read a potential for interest and, thus, a promising sales opportunity.

The disconnect between what is appropriate and what is profitable on the part of algorithms is an instance of what Jed Brubaker calls "sensitive algorithmic encounters." Brubaker explores "how people navigate encounters with algorithmic content related to the death of friends and family" in order to argue for the importance of understanding the broader social context in which algorithms operate. Additionally, Brubaker and Jiang have written extensively on how machine learning algorithms can be deployed in the context of social media in order to make the algorithm more appropriate in the context of death and mourning.[6] Profit, however, not social sensitivity continues to drive the social media algorithms. The algorithms became smarter. They were taught by humans that *death* rather than *bleach* can be the key element to be exploited in creating the conditions for consumption: in other words, discourses around death generate increased user activity and are worthy of pursuing. Consumerism here was to be expressed in the consumption of media itself rather than in the products advertised through media. Don't buy bleach. Just keep watching, clicking, posting, liking.

In researching Todd's story, I stumbled on the *ABC News* article "Bullied Teen Leaves behind Chilling YouTube Video" by Christina Ng published on October 12, 2012.[7] The online news article featured a prominent headline, followed by a video player, which featured ads followed by hardwired stories about death and suicide. Advertising content was labeled as "Ad 1 of 1." The online news stories themselves were seen as the default content and did not have an additional label. The website's video ranking system, similarly to the *YouTube* channel, proceeded to center on suicide.

Here, the clustering of thematic content was done through human curation. Its relatively hardwired architecture makes it easier to detect how its autoplay feature was constructed. In the context of Web 2.0 social media, the code is made legible, and the logic of selecting *what* to display is rendered visible. The video sequence went as follows: commercial, excerpt from Todd's video, commercial, "Bullied Canadian Teen Found Dead at Home," commercial, "Oklahoma Teen Shot Himself at Junior High School," commercial, "Teen Kills Herself after Being Bullied: Family," commercial, "Bullying Suspect in N.Y. Teen's Suicide," commercial, "Phoebe Prince's Mom Confronts Bullies." The sequence played in the same order every time I visited the website from my profile or from an Incognito Window (where browsing history is supposedly ignored). Here, the stories were numerically represented, and their sequence was hardwired as follows: 17466390, 17339232, 15885587, 15287850, and 13533930. A glimpse at the source code of this website sheds light on the mechanism behind the autopay sequence on suicide.

```
window.abcnvideo = window.abcnvideo || {};
window.abcnvideo.storysection = "international";
window.abcnvideo.storyid = "17463266";
....
window.abcnvideo.playlist = [17466390, 17339232,
  15885587, 15287850, 13533930];
```

What is interesting here is that the playlist selection appeared to be carefully curated through human intervention, at least through human editorial tagging of other *ABC News* stories from that time period that engage with the keywords: bullying, suicide, video. This tagging was evident through a closer look at the metadata behind the html page:

```
<meta name="news_keywords" content="Amanda Todd, Port Coquitlam,
  Vancouver, Canada, teen, suicide, bullying, YouTube, video" />
```

This particular website is an example of what Web 2.0 looked like and how it functioned in 2012. In a Web 2.0 context, documents were linked through personalization and through likeness.[8] Here, metadata played a crucial role in organizing similar "documents" or "stories"—a phenomenon that would shift with the organization of data in the context of Web 3.0. More specifically, the meta-name "suicide" was used to cluster written news stories and online video content: "Oklahoma Teen Shot Himself at Junior High School" is a story that ran on the *ABC News* website on September 27, 2012, "Teen Kills Herself after Being Bullied: Family" a story from March 9, 2012, "Bullying Suspect in N.Y. Teen's Suicide" a story from January 4, 2012, and "Phoebe Prince's Mom Confronts Bullies" a story from May 5, 2011. Representing the span of a year and a half, these stories about teens, bullying, and suicide were given unique ids and were then linked and sequenced in a cohesive playlist.

The *ABC News* playlist exemplifies the state of "linked documents" in the early 2010s. Documents were linked because of their "sameness" or "likeness." The tagging or metadata of the video content allowed for subsequent thematic pulls. In other words, tagging different video "assets" with standard keywords such as "bullying" or "suicide" was a first step toward the ability of either human or machine sorting on the basis of likeness or similarity. Looking at this intermediate step in the development of digital media reveals the ways *sameness* was made possible both by the conceptualization of documents as linked on the basis of relatedness through both editorial selection and by the database logic of clustering.

This linking of related documents was also explicitly present in the 2012 version of the *YouTube* platform. The *WayBack Machine* preserved a capture of the *YouTube* screen from October 13, 2012, a month after Amanda Todd's video went live and just days after her suicide. The *WayBack Machine* allows us a glimpse into the early stages of algorithmically driven homogeneity. The source code from the capture reveals not hardwired lists but, rather, content generated based on a user (via a session ID), clustering around the vague requirement of being "related" via the code *"feature=related."* The mechanisms for establishing this "relatedness" were obscured. We were left with the accurate results of "more of the same." *YouTube*, at this moment, was forecasting the new web, Web 3.0 in which linked data and algorithms present information across media based on sameness–a step beyond personalization. This format of information distribution has come to be known as the "filter bubble."

Amanda Todd's video illuminates the multiple ways death, suicide, and mourning were harnessed in social media for capital gain. Banner ads for clearing product *Tide* on *YouTube* and heavy advertising on the *ABC News* website speak to ways in which sensitive content was deemed as highly profitable content because it drew broad interest and thus resulted in increased online participation. Sensitive content was aggregated through sameness in order to hold the viewers' attention longer, hence the ability to deliver more advertising. This logic of "more-of-the-same equals more-advertising-revenue," once curated by an editorial team, became articulated through Web 3.0 media algorithms of sameness. Moving beyond death as document to death as data, it is important to address what algorithms do and more importantly *how* they create filter bubbles of death.

MULTIPLE ALGORITHMS

In *Hello World: Being Human in the Age of Algorithms*, Hannah Fry offers a succinct, yet highly adaptable, definition of an algorithm: "The invisible pieces of code that form the gears and cogs of the modern machine."[9] Taina Bucher makes a valuable addition: the code in question is a sequence of commands: an algorithm is a "set of instructions for solving a problem or completing a task following a carefully sequential order."[10] However, the algorithms we are currently encountering have homogeneity as their objective. Sameness is constructed through four types of algorithms. Fry classifies algorithms based on function in four major categories: (1) prioritization or the making of an ordered list (*Google Search*, *Netflix*); (2) classification or the picking of a category (advertising and measurable types); (3) association or the finding of links (dating algorithms, *Amazon*'s recommendations); (4) filtering or the isolating of what's important (speech recognition, facial recognition, and other processes of separating signal from noise).[11] Further, Fry argues that based on paradigms, the algorithms can be divided into two main groups: (1) rule-based algorithms where instructions are constructed by a human and are direct and unambiguous (the logic of a mac n' cheese recipe) and (2) machine learning algorithms which are inspired by how living creatures learn.[12] Algorithms thus vary in their logic and purpose. They also work together to create larger filtering structures.

As Bucher writes, the *Netflix* queue is produced through a series of algorithms working together:

> To give an impression of the many algorithms that play into designing this overall Netflix experience, Gomez-Uribe and Hunt list at least eight different algorithms, including the personalized video ranker (PVR), which orders the entire catalog of videos for each member in a personalized way; the Top-N Video ranker, which produces the recommendation in the Top Picks; and the page generation algorithm, which works to construct every single page of recommendations.[13]

While Bucher argues that algorithms are constantly changing and are being fine-tuned, I suggest that their core end goal, their problem remains constant. In other words, while the code is subject to variability, the general outcome is not. In shoring up "likeness" or "sameness," algorithms, such as those deployed by *Netflix, YouTube*, *Facebook,* and *Apple Music*, work together, feeding into one other the different bits and pieces of the "if . . . then" programmatic logic that pushes aside data points considered to be different. "If . . . then" is a logic that enforces the exclusion of the "Other."

While the *Netflix* queue and the *Facebook* feed have been seen as examples of an efficient and prized algorithm, both have been subject to much debate about the intervention and bias of humans into what would be considered a neutral and unbiased algorithm. Bucher highlights the employment of journalists and moderators in *Facebook*'s seemingly neutral feed and in taggers of the *Netflix* queue.

> Although Facebook's news feed is often heralded as a prime example of algorithmic finesse, the feed is far from simply automatic or without human intervention. As with the trending section, Facebook enlists human intervention of the news feed as well. . . . Netflix, for example, employs a wide range of "taggers," whose responsibility it is to assess the genre, tone, and style of a film's content to help determine what users might want to watch next.[14]

Bucher suggests that algorithms and humans are implicated in "assemblage or coming together of entities [which become] more or less human, more or less nonhuman."[15] The algorithmic process is thus seen as situated on a continuum between human and nonhuman agency. It is important to note that human agency prior to 2016 was most often associated with the introduction of bias. As the case of *Facebook* illustrates, the nonhuman algorithms are seen as better, more accurate, and less biased.

In the face of widespread accusations against the subjective biases of the human editors, *Facebook* decided to fire the thirty-six journalism graduates contracted to write and edit the short descriptions for the trending topic module. In a bid to reduce bias, Facebook announced instead that they would replace them with robots.[16] Algorithms, thus, are seen as pure when human agency and, hence, bias are curtailed. Nonhuman algorithms are seen as more precise.

Both, however, perform the same function in relation to linked data: they create homogeneity reflected back at a user, a phenomenon I call *solipsistic homogeneity*. In filtering and reifying more of the same, they both insist on and institute sameness based on data or information they already have. This process thus operates on the logic of solipsism. Further, they filter out difference and thus both value and uphold homogeneity.

THE FILTER BUBBLE IS UNFORGIVING

The algorithmic rational and seemingly impartial calculations create a world in which we get more of what we supposedly *like* instead of at least a bit of what we might sensibly *need*. The algorithms of mourning structure a "filter bubble" that is profoundly devoid of empathy. The condition of receiving more of the same has been theorized at length by Eli Pariser, who coined the term "filter bubble" as "a unique universe of information for each of us."[17] Filter bubbles personalize and customize news and other experiences based on what is of "relevance" to the user. The reason for this aggregation of information based on perceived personal relevance, as Pariser argues, is commercial. Pariser writes that "as a business strategy, the Internet giant's formula is simple: The more personally relevant their information offerings are, the more ads they will sell, and the more likely you are to buy the products that they're offering.[18] This personalization has been found problematic because "filter bubbles created by the platform's algorithms—curated echo chambers that trap users in a world of news that only affirms their beliefs.[19]" Filter bubbles thus share the following properties: they are algorithmically sorted, they are rooted in linked data that incorporates both media as data and viewers as data, and they span across media platforms. Pariser articulates three important effects on humans, on those of us, on all of us, subjected to corporate-driven filter bubbles. First, he argues, "you're alone in it."[20] Second, he suggests, "the filter bubble is invisible" as "from within the bubble, it's nearly impossible to see how biased it is."[21] Third, Pariser warns that filters are really difficult to avoid. Crucial to the creation and maintenance of the filter bubble are both the linked data to be presented and the sorting algorithms that establish relevance.

While the operation of the filter bubble seemed more explicit on *YouTube*'s video queue, it was more obscure and harder to detect platforms like *Facebook*. Taina Bucher has detailed the history and development of the *Facebook* news feed. Launched in 2006, the news feed is one of *Facebook*'s primary and most successful features. The news feed, placed in the middle of a user's home page, is a continuous stream of updates serving a user with stories of what their friends have been up to: "In its twelve years of existence, the news feed has gone from being a single feed of reverse-chronologically-ordered updates to becoming fully edited and governed by machine learning algorithms."[22]

The machine learning algorithm, also known as the Edge Rank, as described by Bucher, relies on ranking mechanism of relevancy or likeliness of continuous interaction: "Once the relevancy score for each post has been determined, a sorting algorithm can then put them in the right order for the user to see."[23] The posts weighted and sorted here are user-generated as well as paid advertising content. The more a user interacts with a particular type of content, the more similar user-generated and paid content is delivered.

If the filter bubble of 2012 delivered ads about cleaning products, the algorithm makers learned from their mistake in linking bleach to detergent in the Todd case and, subsequently, adjusted the algorithm so that it more accurately determined that a keyword for ad revenue can be "suicide" itself. Corporate media giants commodified and attempted to capitalize on "suicide" and "death" themselves. As long as users keep interacting, the content will keep coming. Experiences such as "death" and "suicide" are commodified and algorithmically delivered to unsuspecting viewers in the form of filter bubbles. Visiting Todd's video on *YouTube* in 2019 did not result in detergent ads. Instead, it produced a ranking of other videos dealing with suicide that automatically played, which is, nonetheless equally jarring: "Interested in suicide? Check our other content on the same issue" screams the *YouTube* page. With "AUTOPLAY" automatically enabled, *YouTube* created an "Up next" playlist for me featuring "Cyberbully suicide attempt scene Sia—breathe me" that had had 4.4 million views, followed by "Stunts that Sadly Cost People Their Lives" with 1.4 million views, then "Stalking Amanda Todd: The Man in the Shadow" with 6.9 million views, a video about kids in prison in fourth place, and coming up fifth in my feed, a Ted talk on "The Bridge between suicide and life" by Kevin Briggs with 2.6 million views. The algorithm had found more of the same. Notable here is that on the top of my most relevant, autoplay options was a suicide attempt scene. So, what are we to make of these homogeneous streams of social media centering around topics of suicide and death?

ALGORITHM OF MOURNING

In thinking about insensitive algorithmic encounters, I turn to a discussion of the *implications* of the solipsistic algorithm of mourning. In February 2019, a member of my community committed suicide and friends and family were actively posting to his *Facebook* page as well as sharing photographs and notes in his memory. Our community was mourning. We were tagging, liking, following, and clicking online. The algorithm did not fail to capitalize on this newly found hot topic. Soon my *Facebook* news feed resembled an obituary page as it became populated not only by posts by friends but also by paid content from news sources I usually read such as *NPR*, and the *New York Times*. These news feed items focused on suicide: "Applied Suicide Intervention Skills Training," "Deaths from Drug and Suicide Reach a Record in the U.S.," "Everyone Around You is Grieving. Go Easy," "The poet W.S. Merwin has died at 91," "Alan Krueger, an economist known for his work on minimum wage and employment, is dead at 58," "Jake Phelp Dies at 56," "Okwui Enwzor, Curator Who Remapped Art World, Dies at 55," "Sydney Aiello, a student who survived the deadly shooting in Parkland, Florida, last year has died by suicide, her family said this week." How many death notices can one take in a month of mourning?

The stakes of *solipsistic homogeneity* in the context of algorithms of mourning are explicitly biopolitical. By focusing on sameness, algorithms of mourning inflict autoimmunitary violence on those who are subjected to their logic. Here, I am evoking Jacques Derrida's notion of the autoimmunitary process, which he defines as "that strange behavior where a living being, in quasi-suicidal fashion, "itself" works to destroy its own protection, to immunize itself against its 'own' immunity."[24] To surround people who are mourning the suicide of a community member with an overwhelming amount of information about suicide and death for the purpose of commercial profit is to make the work of the algorithm doubly quasi-suicidal. The doubling here addresses the immunitary functions of mourning in society on the one hand, and of securing continuous engagement and participation on the part of the algorithm on the other. First, it complicates the immunitary processing of mourning a loss. This is an important observation, because the loss is a loss within and never external to humanity, to community, to friendship. More broadly, however, it asserts a conversation about suicide in a single, overwhelming, dominant direction and unfortunately leads to the literal intensification of suicide rates via the contagion effect, as seen in the study of the rise of teen suicide following *Netflix*'s release of *13 Reasons Why*. The commercially driven suicide filter bubble, which then displays more of the same media in queue here, visualizes the autoimmunitary aggression that algorithms inflict. A second way algorithms of mourning act in quasi-suicidal fashion can be observed in the ways in which they turn against the system of popularity and social media participation that makes them possible. In other words, the primary purposes of media algorithms are prolonged engagement, more social activity, and media consumption. When mourning became commodified and algorithmically omnisciently delivered, the act of mediated interaction itself comes under symbolic and literal threat: audiences choose to retreat from their feeds and audiences face higher risks in retreating from the world.

I name the algorithm that articulates likeness and relevance in the context of death an *algorithm of mourning*. This algorithm presents content related to death and suicide to someone who has viewed, liked, or in other ways interacted with content that engages with death and suicide and by virtue in this action could be considered the mourner of the death. It evokes the inhumanity of algorithmic logic through two important points: first, the algorithm does not understand mourning, or is unable to mourn, and second, the algorithm can be accused of perpetuating death and thus extending the scope of those in mourning. Why mourning? In viewing content related to death or suicide, I argue that one is either explicitly or implicitly involved in the act of mourning. I situate mourning here not as much as an individual act but rather as a community act. In his poignant text *The Work of Mourning*, Jacques Derrida captures the complexity and force of mourning in a letter to Francine Loreau:

> This being at a loss says something, of course, about mourning and about its truth, the impossible mourning that nonetheless remains at work, endlessly hollowing out the depths of our memories, beneath their great beaches and beneath each grain of sand, beneath the phenomenal or public scope of our destiny and behind the fleeting, inapparent moments, those without archive and without words (a meeting in a cafe, a letter eagerly torn open, a burst of laughter revealing the teeth, a tone of the voice, an intonation on the phone, a style of handwriting in a letter, a parting in a train station, and each time we say that we do not know that we do not know if and when and where we will meet again). This being at a loss also has to do with a duty: to let the friend speak, to turn speech over to him, his speech, and especially not to take it from him, not to take it in his place—no offense seems worse at the death of a friend (and I already feel that I have fallen prey to it)—to allow him to speak, to occupy his silence or to take up speech oneself only in order, if this is possible, to give it back to him.[25]

Mourning calls on memory and destiny, on loss and duty, on silence. Mourning is time. It is in time. Takes time. It allows us to reflect on our time together and our time apart. Mourning is the act of "let[ing] the friend speak" rather than speaking for a friend. Here Derrida makes a radical point. A humane point of empathy. Mourning is to be *decentered* from the mourner. Mourning is an *act oriented toward* a friend. It is about them. It should be about them. It is his silence, his voice. To which I add, her silence, her voice; their silence, their voice. To mourn is to turn to another and to revel in the loss, in the void, and not to take over the absence and make ourselves the center of the act.

The algorithm of mourning fails at capturing precisely this selfless decentering that mourning evokes. Instead, it does the opposite: it situates us in the void, in the space of loss. The algorithm centers us and creates a new bubble around us, that is about us. The *solipsistic homogeneity* that it instills erases the other. The other disappears. We take center stage. And Derrida pleaded, "especially not to take it [the speech, the voice] in his place."[26] Based on linked data and a logic of sameness, the algorithm of mourning orients toward continued consumption of content related to suicide and death because of our actions. It centers on us and creates a homogeneous view of what we are perceived to "like." The inhuman algorithm of mourning offers no other, no silence, no comfort. It exploits a vulnerability in an attempt to capitalize. It aims to bring profit to the media giants rather than to benefit a distraught audience.

The second point of perpetuating death and thus extending the scope of those in mourning can be seen in the case of the removal of the contested suicide scene from Netflix's show as signaled by the National Institute of Mental Health. The scene was linked to research linking the scene to increased suicide rates in real life. The *Netflix* show "13 Reasons Why" was associated with a 28.9 percent increase in suicide rates among U.S. youth aged ten to seventeen in the month (April 2017) following the show's release, after accounting for ongoing trends in suicide rates, according to a study published in Journal of the American Academy of Child and Adolescent Psychiatry. The findings highlight the necessity of using best practices when portraying suicide in popular entertainment and in the media.[27]

It took *Netflix* two years to delete this scene.[28] The show creator Brian Yorkey expressed on *Twitter* his intention about creating the scene and reasons for removing it.

> It was our hope, in making 13 Reasons Why into a television show, to tell a story that would help young viewers feel seen and heard, and encourage empathy in all who viewed it, as much as the bestselling book did before us. Our creative intent in portraying the ugly, painful reality of suicide in such graphic

> detail in Season 1 was to tell the truth about such an act, to make sure no one would ever wish to emulate it. But as we ready to launch Season 3, we have heard concerns about the scene from Dr. Christine Moutier at the American Foundation for Suicide Prevention and others, and have agreed with Netflix to re-edit it. No one scene is more important than the life of the show, and its message that we must take better care of each other. We believe this edit will help the show do the most good for the most people while mitigating any risk for especially vulnerable young viewers.[29]

Let's unpack this statement: "No one scene is more important than the life of the show." Given the biopolitical concerns here over the life and death of the spectator, of the witness, the show's creator expresses concern here with the life of the show. The show must go on first and foremost. What about Hannah Baker? What about the life of the "viewers." Viewers? The term here is deeply problematic. It is deeply and uncritically embedded in a framework of media consumption—framework that, in this case, has been considered deadly. The "viewers" are people. Humans. Friends. Friends of Hannah. Friends to each other. It is the human life rather than the media life that we should bring back into focus. Human death, rather than media death should be of concern here.

I further want to situate the show and this particular scene back into the consumerist algorithmic logic of *Netflix*. It is not just about the scene. The conversation needs to center around the bubble a scene like this creates. Expressing interest in the *13 Reasons Why* episode already guaranteed a change of content relevant to the user. After watching the above mentioned episode, related films that appeared at my queue included titles such as *My Suicide*, *Face2Face*, and *Not Alone*. Similarly, I previously related how my recent viewing of Amanda Todd's story on *YouTub*e generated a suicide theme for my autoplay video queue. Content that centers around suicide and death and creates a filter bubble driven by the inhumane algorithm of mourning. The potential deadly effect of media portrayals of suicide and death has a name: "the suicide contagion effect." Elahe Izadi highlights the key aspects of this effect. The contagion effect can be summed up as follows: "The amount, duration and prominence of suicide news coverage 'can increase the likelihood of suicide in vulnerable individuals.'"[30] The phenomenon of contagion via fictional media has been challenged by scholars who argue that the current research does not support such correlation.[31] The critique, however, engages with studies that examine the effect of a singular media text on the moods and perceptions of its viewers. I want to highlight, however, the importance of questions of "amount, duration, and prominence." Given the serial and homogeneous structure produced by media algorithms, I suggest that the filter bubble created by the inhumane algorithm of mourning can be seen as a prime example of the contagion effect and would merit in depth psychological studies.

The inhumane algorithm of mourning, I argue, inflicts autoimmunitary violence. The autoimmunitary process is doubly suicidal as it features two suicides. Speaking of the hijackers in the terrorist attack of September 11, Derrida writes of the two suicides that one is their own (and one will remain forever defenseless in the face of a suicidal, autoimmunitary aggression—and that is what terrorizes the most) but also the suicide of those who welcomed, armed, and trained them.[32] The double suicide is thus of the agent and of the network that created the agent in the first place. The doubly suicidal or autoimmunitary violence comes to the surface in the case of *13 Reasons Why*. The show had to cut out of its fabric to kill the scene that made it popular that made it live. It had to self-mutilate because what was subjected otherwise was the public that made it live. The scene and its filter bubble exemplify the ways the inhumane algorithm of mourning is caught in the biopolitical double bind of autoimmunitary violence: it makes live by increasing viewership and threatens to kill because it is simultaneously promoting death in the viewership. In other words, if the algorithm does not perform its job, the media object is left unconsumed and thus dies off by being unpopular. By becoming popular, however, media, driven by the algorithm, is forecasted to die off as well because its fame creates a contagion effect and threatens to render its consumers, its viewers, deceased.

THE VALUE OF DIFFERENCE

To recognize mourning as a human and humane act is to elevate the importance of difference and empathy. Mourning emerges as a potential potent antidote to solipsism, to homogeneity, as it invites us to acknowledge an other, to let speak, to listen. In exposing and defying the algorithmic logic of sameness, we need to become active seekers of difference. In the age of algorithmic sameness, it is up to us preserve our capacity to *orient ourselves to another*. Our humanity depends on it.

NOTES

1. Margaret Gibson. "Automatic and automated mourning: messengers of death and messages from the dead." *Continuum: Journal of Media and Cultural Studies*, 2015, Vol. 29, No. 3, 340.
2. Ibid., 341.
3. Nikolaus Lehner. "The work of the digital undead: digital capitalism and the suspension of communicative death." *Continuum: Journal of Media and Cultural Studies*, 2019, Vol. 33, No. 4, 476.

4. Ibid., 478.
5. Ibid., 484.
6. Jed Brubaker. "Socializing algorithms." *Trustworthy-Algorithms.org*. http://trustworthy-algorithms.org/whitepapers/Jed%20Brubaker.pdf
7. ABC News. https://abcnews.go.com/International/bullied-teen-amanda-todd-leaves-chilling-youtube-video/story?id=17463266
8. Tim O'Riley. "What is Web 2.0? Design Patterns and Business Models for the Next Generation of Software." *OReily.com*, 2005.
9. Hannah Fry. *Hello World: Being Human in the Age of Algorithms*. New York: W.W. Norton & Company, 2019.
10. Taina Bucher. *If . . . Then: Algorithmic Power and Politics*. London/New York: Oxford University Press, 2018, 20.
11. Ibid., 8–9.
12. Ibid., 10–11.
13. Ibid., 47–48.
14. Ibid., 53–54.
15. Ibid., 54.
16. Ibid., 55.
17. Eli Pariser. *The Filter Bubble: What the Internet Is Hiding from You*. New York: The Penguin Press, 2011, 9.
18. Ibid., 7.
19. Ben Sullivan. "'I Just Got Really Mad': The Norwegian editor tackling Facebook on censorship." *Vice*, 2016.
20. Pariser, *The Filter*, 9.
21. Ibid., 10.
22. Bucher, "If . . . Then," 74.
23. Ibid., 78.
24. Jacques Derrida. "Autoimmunity: Real and symbolic suicides." *Philosophy in a Time of Terror: Dialogues with Jürgen Habermas and Jacques Derrida* [translated by Giovanni Boradorri]. Chicago: University of Chicago Press, 2003, 94.
25. Jacques Derrida. *The Work of Mourning* [translated and edited by Pascale-Anne Brault and Michael Naas]. Chicago/London: Chicago University Press, 2001, 94–95.
26. Ibid., 95.
27. National Institute for Mental Health. "Release of '13 Reasons Why' Associated with Increase in Youth Suicide Rates." *National Institute for Mental Health*, 2019.
28. Allyson Chiu. "A graphic suicide scene in '13 Reasons Why' drew outcry. Two years later, Netflix deleted it." *The Washington Post*, 2019.
29. Yorkey Brian. "A statement from our show creator Brian Yorkey." 13ReasonsWhy. *Twitter,* 2019.
30. Elahi Izadi. "What happens when a suicide is highly publicized in the wrong way: The suicide contagion effect." *Washington Post*, 2014.
31. Christopher Ferguson. "13 reasons why not: A methodological and meta-analytics review of evidence regarding suicide contagion by fictional media." *Suicide and Life-Threatening Behavior*, 2019, vol. 49, No. 4: 1178–1186.
32. Ibid., 95.

BIBLIOGRAPHY

Brubaker, Jed. "Socializing algorithms." *Trustworthy-Algorithms.org*. http://trustworthy-algorithms.org/whitepapers/Jed%20Brubaker.pdf

Bucher, Taina. *If . . . Then: Algorithmic Power and Politics*. London/New York: Oxford University Press, 2018.

Chiu, Allyson. "A graphic suicide scene in '13 Reasons Why' drew outcry. Two years later, Netflix deleted it." *The Washington Post*, 2019. https://www.washingtonpost.com/nation/2019/07/16/reasons-why-suicide-scene-pulled-netflix/ (Accessed September 2, 2019).

Dean, Michelle. "The story of Amanda Todd." *The New Yorker*, 2012. https://www.newyorker.com/culture/culture-desk/the-story-of-amanda-todd (Accessed September 16, 2019).

Derrida, Jacques. *The Work of Mourning* [translated and edited by Pascale- Anne Brault and Michael Naas]. Chicago/London: Chicago University Press, 2001.

Derrida, Jacques. "Autoimmunity: Real and symbolic suicides." *Philosophy in a Time of Terror: Dialogues with Jürgen Habermas and Jacques Derrida* [translated by Giovanni Boradorri]. Chicago: University of Chicago Press, 2003, pp. 85–136.

Ferguson, Christopher. "13 reasons why not: A methodological and meta-analytics review of evidence regarding suicide contagion by fictional media." *Suicide and Life-Threatening Behavior* 49(4), 2019, 1178–1185. doi:10.1111/sltb.12517.

Fry, Hannah. *Hello World: Being Human in the Age of Algorithms*. New York: W.W. Norton & Company, 2019.

Gibson, Margaret. "Automatic and automated mourning: messengers of death and messages from the dead." *Continuum: Journal of Media and Cultural Studies* 29(3), 2015, 339–353. doi:10.1080/10304312.2015.1025369.

Izadi, Elahi. "What happens when a suicide is highly publicized in the wrong way: The suicide contagion effect." *Washington Post*, 2014. https://www.washingtonpost.com/news/to-your-health/wp/2014/08/12/what-happens-when-a-suicide-is-highly-publicized-in-the-wrong-way-the-suicide-contagion-effect/ (Accessed September 3, 2019).

Jialun "Aaron" Jiang, and Jed R. Brubaker. 2018. Tending unmarked graves: Classification of post-mortem content on social media. *Proceedings of the ACM on Human Computer Interaction* 2, CSCW, Article 81 (2018), 19 pages. doi:10.1145/3274350.

Lehner, Nikolaus. "The work of the digital undead: digital capitalism and the suspension of communicative death." *Continuum: Journal of Media and Cultural Studies* 33(4), 2019, 475–488. doi:10.1080/10304312.2019.1627289.

National Institute for Mental Health. "Release of '13 Reasons Why' Associated with increase in youth suicide rates." *National Institute for Mental Health*, 2019. https://www.nimh.nih.gov/news/science-news/2019/release-of-13-reasons-why-associated-with-increase-in-youth-suicide-rates.shtml (Accessed September 15, 2019).

Ng, Christina. "Bullied teen leaves behind chilling youtube video." *ABC News*, 2012. https://abcnews.go.com/International/bullied-teen-amanda-todd-leaves-chilling-youtube-video/story?id=17463266 (Accessed September 20, 2019).

O'Riley, Tim. "What is web 2.0? Design patterns and business models for the next generation of software." *OReily.com*, 2005. https://www.oreilly.com/pub/a/web2/archive/what-is-web-20.html (Accessed September 4, 2019).

Pariser, Eli. *The Filter Bubble: What the Internet Is Hiding from You*. New York: The Penguin Press, 2011.

Sullivan, Ben. "'I just got really mad': The Norwegian editor tackling facebook on censorship." *Vice*, 2016. https://www.vice.com/en_us/article/aeknxg/i-just-got-really-mad-the-norwegian-editor-tackling-facebook-on-censorship-aftenposten (Accessed September 5, 2019).

Yorkey, Brian. "A statement from our show creator Brian Yorkey." 13ReasonsWhy. *Twitter*, 2019. https://twitter.com/13ReasonsWhy/status/1150987786243018752.

Chapter 7

Algorithms, Identity, and Cultural Consequences of Genetic Profiles

Amanda K. Girard

When my brother received an *AncestryDNA* test kit for Christmas, he was excited to learn who he was beyond exaggerated family stories. We had always been told that we were part Native American. He wanted to confirm that he was "enough" Native American to qualify for certain scholarships when applying to law school. While my brother's motivation was instrumental it also relates to other stories that show genetic proof of ancestry as meaningless to historically established cultures.

Recently, Sen. Elizabeth Warren made the news for a similar misunderstanding of Native American heritage. Warren, motivated by Donald Trump's taunts and family lore, used a DNA testing to show that she had indigenous ancestry.[1] However, genetic testing does not really prove or disprove race. While Warren was attempting to identify as a Native American, her claims and DNA proof only served to undermine hard earned tribal rights and benefits that have a long history of exploitation and possible removal.[2] Further, most people, like my brother, do not know the steps necessary to confirm tribal citizenship and that this process is different for each tribe. Identifying as Native American means much more than checking a box identifying race or ethnicity on a form—even if that form asks one to self-identify. Regardless, my brother's results showed information about being Northern European and little else. I was surprised.

While there are plenty of gifted storytellers in my family, as a researcher, I questioned my brother's results. I started digging through the websites for information on companies like *AncestryDNA* and *23andMe* and reading what others have found out about them. Initially, I was unable to find specific information about how these companies process DNA samples, but I did find out that some companies outsource that work to specialized labs. I also found a lot of older popular articles on the inaccuracy of some test results and an explanatory link to the algorithms and databases these companies used and did not share with the public.[3]

The specifics on DNA processing for Native American heritage were different for different sites and all had disclaimers. Reading the FAQ for *Cellular Research Institute Genetics (CRI)*[4] yielded an explanation for why my brother's results may be inaccurate, unable to demonstrate Native American heritage where it may in fact exist. In response to the frequently asked question, "Can CRI confirm my race?" they answer, basically, no. Further, they explicitly address the circumstance that many "users" want to confirm Native American heritage and explain that their results do not constitute legal proof of "Biogeographical Ancestry." Legally, each Native American tribe has a set of guidelines to establish citizenship, and these conditions vary from tribe to tribe. The *CRI* test determines your "Biogeographical Ancestry," but defined as *where* your evolution occurred based on your genetic makeup, not as your race or ethnicity. Chomsky notes that none of these sites use the term "race:" instead they often rely on "ethnicity," a word previously used to indicate culture and identity and now considered relative to genetic measurement.[5]

CRI argues that race and ethnicity are inadequate terms to define the science and history they engage in. They explain: due to "the complex anthropological history of human migration," it is possible "to have some ancestry from other haplogroups around the globe." In other words, although your genetic information suggests where you are from that may have been so many generations ago, and humans have moved around so much that a company that processes your DNA cannot confirm or deny your exact placement within a group. Similarly, *23andMe* claims that "Native American ancestry tends to be five or more generations back, so that little DNA evidence of this heritage remains."[6] *AncestryDNA*[7] also makes it clear that the test is not legal proof of Native American ancestry. *AncestryDNA* explains that indigenous DNA can be lost based on the randomness of inheritance. Chomsky sees these claims as furthering a historical colonizer agenda to rid America of its indigenous people partially through biological dismissal.[8] She sees mail-in genetic testing as "conflating ethnicity with geography, and geography with genetic markers." So why take the test?

While genetic testing has become readily available to the public through companies like *23andMe* and *AncestryDNA*, their results are based on databases and algorithms the companies will not release to the public. In spite of multiple reports of inaccuracy when evaluating similar or even the same data have made headlines, these companies continue to successfully sell the idea of easy genetic testing to provide a historic profile of identity. In this process, incomplete results and the potential lack of recognition of certain groups make these tests an inappropriate measure of identity. The resulting marginalization and erasing of populations have far reaching cultural ramifications. Measures that have been put in place to recognize and recover limited populations may not be used and those groups could eventually officially vanish.

While the different sites' Frequently Asked Questions answered my originally query about race, I continued searching other sites to reveal similarly evocative information. For example, *AncestryDNA* uses the term "ethnicity estimate" to describe the results of its testing, further exhibiting that companies selling these kits do their best to avoid using the term "race" at all costs. Moreover, ethnicity *prediction* is not new. *Diversity Inc.*, for example, predicts ethnicity using first and last names as data.[9] *AncestryDNA* estimates by "[c]reating an ethnicity estimate based on your DNA sample is a complex process based on probability, statistics, shared DNA, and ongoing research and science." How, I wonder, are these DNA tests processed and analyzed? What are *estimates* based on? Further, given their biases, why are these tests so appealing to the public? Finally, do the results change users' views of themselves and their individual identity?

To address these questions, I consider how the tests developed, the role of users' information in this technology, and the geolocational cultural aspects of identification. In doing so, I answer the call for "interdisciplinary research and scholarship in information studies and library and information science that intersects with gender and women's studies, Black/African American studies, media studies, and communications to better describe and understand how algorithmically driven platforms are situated in intersectional sociohistorical contexts and embedded within social relations" as outlined in Safiya Umaja Noble's *Algorithms of Oppression.*[10] Overall, the drive to know oneself leads to the success of genetic testing. Sequoya Yiauek believes that genetic testing "asks the question of who we are head on," an essential question, but also one that may come with disappointing answers.[11] Yiauek was persuaded by family to understand individual ancestry better. He adapted his life to a narrative of heritage that his results did not match. His story shows that genetic testing can have long-lasting ramifications for identity. While, they promise a scientific knowledge of self.

These mail-in testing kits began as cheap and easy ways to gain medical information.[12] While some are still marketed to help understand predispositions to certain medical conditions or allergies, most quickly and quietly switched to understanding where potential family came from. This shift can be accounted for in several ways. Because the information processing on ancestry involves predictions, not medically defined data, genetic testing for ancestry escapes regulation. The Federal Drug Administration regulates direct-to-consumer tests as medical devices to assure validity. However, tests for "non-medical, general wellness, or low-risk medical purposes are not reviewed by the FDA before they are offered," according to the FDA website.[13] Regulations regarding DNA testing have not kept up with the new technologies available.

The ethics of expediency, or the development and implementation of technology taking priority over moral consideration, is undoubtedly at work within these quick test companies.[14] Consequently, more than a little bit of the ethos, "Move Fast and Break Things," *Facebook*'s first motto, contributes to expediency here. In her book, *Race against Technology,* Ruha Benjamin asks, "What about the people and places broken in the process?"[15] Who, we must ask, is this genetic testing hurting and who is it helping? Yiaueki was hurt. In his experience he discovered that DNA can reveal hidden information—sometimes meant to be lost—and come with "real risk."[16] Because our legal system has not kept pace with technology, it is ill-prepared to address these concerns. In fact, in order to sidestep any potential lawsuits, many of the popular DNA tests have disclaimers about the information that they offer—especially medical information. For the purpose of this chapter, however, it is the ancestry testing and analysis that mail-in DNA companies offer that I address.

Most of the available tests work with DNA samples using three genetic testing methods: autosomal DNA, Y-DNA test, and mitochondrial DNA testing.[17] Each test type has different limitations. An autosomal DNA test can identify relatives on both the maternal and the paternal side for five to seven generations. The Y-DNA test can go back to 60,000 years on the paternal side, but only someone with a Y chromosome can take this test. The mitochondrial DNA test can be taken by both men and women and can predict the maternal line for up to 150,000 years. While company websites do not usually directly state what kind of genetic test(s) that they run, many FAQ sections seem to suggest they have been asked. For example, the *CRI* website includes the question, "Can women use CRI genetics?" and answers that they use autosomal testing.[18] Other sites do not offer this information and some, like *AncestryDNA* and *National Geographic*, send DNA to a third party for processing. Consider what a third party lab might get out of this agreement. Beyond being paid by the primary company, they now have your DNA and genetic code. To what end? What are their research goals? What is their privacy policy? And why isn't this information provided to the consumer? There is little doubt that a third party will make some use of the data. Richard Lanham, in *Economics of Attention*, notes that "clean information is not the destiny of humankind" because "clean information is unnatural and unuseful" to us—we wouldn't know what to do with it.[19] *AncestryDNA*, *National Geographic*, and even the third-party testing labs become knowledge workers—"paid for their ability to find, filter, analyze, create, and otherwise manage information" to make it palatable for their users.[20] Who are the users?

Based on a designated user, tests are treated differently. This was demonstrated when a staff writer for *Live Science* sent nine DNA samples to three different companies: *AncestryDNA, National Geographic,* and *23andMe.*[21] Although the writer is male, three of the samples were marked as female for reporting purposes. Both *23andMe* and *National Geographic* required "more personal information" from the subject because there were "unexpected chromosomes.[22]" However, *AncestryDNA* processed the sample marked as "female" without further communication. Letzter's article gives some insight into how companies may be processing samples differently than others. Further, the information one provides to the company also may affect how one's DNA sample is treated. But the consequences of the testing do not end here. Just as one's gender becomes data that may shape the testing, the user's individual genetic data then contributes to building a library of data for further comparison of others' data.

As *AncestryDNA*'s website explains, a customer's DNA sample is compared to a reference panel [in an algorithmic process] in order to find patterns.[23] Once these patterns are determined, an algorithm is used to estimate a range of possibilities to find the most likely regional matches. Like most DNA testing, the more people willing to share their genetic information with the DNA testing brand, the larger the database of information becomes, and therefore, more patterns can be detected. The size of a database is important because information about ancestry is only as good as the amount of information the company already has on file for comparison. In other words, the more people who share their DNA with a company the better predictions the company can make. However, there is no historic database of ancestral DNA; therefore, your DNA is compared to other's contemporary data.[24]

The evidence suggests the use of machine learning algorithms trained on previous data. In theory, genetic data is entered into an algorithm that identifies patterns among the four nucleobases of DNA.[25] Once patterns among users' genes are identified they are compared to the library of genetic data the company has already processed in order to devise whose genetic makeup looks similar or, more valuably, different. Humans only differ by 0.1 percent genetically; geneticists see this difference as race.[26] Finally, the patterns among DNA identified by the algorithm are assigned regional labels. The information about regional labels (British, Thai, etc.) and connection to DNA patterns is tricky. According to a genetic anthropologist, companies often rely on previous research or self-reports to create a reference library for group names.[27] Chomsky points out that this naming "bear(s) an eerie resemblance to the 'races' identified by European scientific racist thinking a century ago."[28] Moreover, the process of creating this library of regional labels is difficult because humans—as was noted earlier—move around, populations identify or are affiliated with different group after wars, and national boundaries change.

Richard Lanham would agree that these companies are acting as designers and not describers. He asserts that "designers make patterns in the physical world, templates for stuff. But when they design information, they are designing nonstuff, templates for how to think about the world, how to act in it."[29] Boundaries are human made. Maps are human made. Labels are assigned to information. And sometimes these labels are "as specific as 'Sardinia'" or as broad as "East Asia."[30] These mail-in genetic ancestry companies are actually creating and assigning deceptively neat labels to sociocultural change throughout the world in order to "reveal" your historic identity. Notably, *23andMe* asserts that the Native American population originally came from northeastern Asia about 15,000 years ago; however, due to war, genocide, disease, and colonization, the "genetic legacy" of these people is now found in Central and South America—further burying the North American Native American history.[31] The labels work rhetorically to name and establish certain areas or people as significant or not. This naming creates metadata. Peter Morville considers this metadata through the lens of information architecture and explains that metadata's purpose is to "encapsulate *aboutness* now to support findability later."[32]

Further, Morville notes that "in an age of location-awareness, when metadata can be attached to people, possessions, and places, the findability and value of our documents and objects will be shaped by strange new forms of sociosemantic *aboutness*"[33] In other words, the metadata attached to a particular person or object will be based on their findability—whether a person or object can be found through location tracking or information available on the Internet or, in this case, as directed by a DNA analysis service. The genetic testing companies create where we can find our ancestry.

While many companies focus on the genetic processing first, *National Geographic* reverse engineered their DNA testing by beginning with research about human migration patterns for up to 200,000 years and then adding the genetic testing component by partnering with Helix or Family Tree DNA.[34] While other companies struggle to justify their geolocational estimates based on the user's genetic data and often update, *National Geographic's Genographic Project* focused on ancient migration patterns more than individual user's DNA and started as a research project with the human DNA kits as part of the public participation component.[35] However, *National Geographic*, like other companies, is still acting as an information filter to match up contemporary metadata about places or regions to genetic patterns produced through algorithms at an outside lab. Drawing on machine learning algorithms and users' saliva, genetic testing affordably integrates medical knowledge and historical and anthropological research about human migration patterns to make labeling and naming feasible, but fallible, with real consequences for the consumer.

Many DNA testing kits now flood the market and a search for the "best" ones reveals quite a few, including *AncestryDNA*, *23andMe*, *CRI Genetics*, *My Heritage*, *Family Tree*, *African Ancestry*, and *Life DNA*. Comparisons of which tests work best are based on what the user is looking for, but *AncestryDNA* wins for having the largest database of fifteen million. *23andMe* is a close second with ten million. However, *My Heritage* may offer the best chance to find global matches, and *Living DNA* is touted as being best for those with heritage in the British Isles.[36] Some services stand out, like *African Ancestry,* for targeting a specific part of the world and researching specific tribes and populations; the costs of such services almost triple the price of other brands.[37] Overall, many services cater almost exclusively to Americans or Western Europeans and are easily found on sites like Amazon or even at Target stores. Those that cater to specific populations, such as a global match or the British Isles, may not have set out to do so. Rather, the DNA they have previously collected has contributed to a unique genetic library, and the algorithms have found more matches within these populations. Chomsky asserts that "few Native Americans have chosen to donate to such databases" siting historical "abuse at the hands of colonial researchers" who profited from Native culture and the general "skepticism toward the notion of offering genetic material for the good of 'science.'"[38] Therefore, Native Americans may be a little found population in any database because the algorithms may not find enough data.

Machine learning algorithms are "trained" to look for certain kinds of data. In this case an algorithm is trained with genotypes, which means that sufficient data has to be entered to verify myriad different possible genetic patterns. It is not surprising that these companies withhold their algorithms from the public. "Secret algorithms" are important parts of many different business models, proprietary, and protected by law.[39] The *CRI Genetics* website highlights that its Chief Scientific Officer, Alexei Fedorov, Ph.D. is a geneticist who created a patented DNA analysis algorithm.[40] While computer modeling has been used in medicine for quite some time, it seems reasonable to question the ethics of an academic geneticist developing an algorithm to process data for profit. But, as in other industries, computational approaches to a wide array of problems are seen as not only good but necessary; and a key feature of cost-cutting measures is the outsourcing of decisions to "smart" machines.[41] Algorithms cut the cost and human power it previously took to analyze strings of DNA. Additionally, "automated systems are alluring because they seem to remove the burden from human gatekeepers, who may be too overworked or too biased to make sound judgments," especially when it comes to matching specific DNA sequences to regions.[42] "Profit maximization, in short, is rebranded as bias minimization. But the outsourcing of human decisions is, at once, the insourcing of coding inequity."[43] Those that develop algorithms that are meant to minimize bias and are seen as neutral technologies still create code based on sociocultural engrained biases.

Creating code to interpret or analyze data is a complex and flawed process. Nonetheless, it is a process that has evolved rapidly. *23andMe* was actually updating its system for interpreting DNA samples while a *Live Science* reporter was comparing results of the same DNA submitted under three different names. While the reporter's initial results showed that he might be a fraction of 1 percent Native American the reassessment showed that he was 100 percent Ashkenazi ancestry.[44] The reporter never believed that he was Native American, but disparity between these assignments of this ancestry makes one wonder how data is being assessed and interpreted. In that same vein, the reporter submitted two DNA samples to *AncestryDNA* and got similar, but not identical results.[45] Despite these stories and the companies' own disclaimers, inaccurate or estimated information does not deter us. We are willing to sacrifice information quality for accessibility and pay for it.[46]

AncestryDNA has actually updated their findings and changed some users' ethnicity estimates as they have updated their database. The company explains that with more DNA samples their reference panels improve and even boasts the ability to break down larger regions now. Are they finding new genetic patterns? Researching previously unrecognized populations? I presume that their algorithm is being updated or better trained as more samples are added. Based on *AncestryDNA's* FAQ page on updates, some users' ethnicity estimates have changed significantly.[47] For example, answers to questions suggest that a user may "lose" a particular ethnicity from a previous result and/or the assignment of a particular region could change drastically. These are not small changes and indicate updates on medical, technological, and anthropological levels.

One wonders if *AncestryDNA* was leaving some groups out of its ethnicity results before—especially those groups who may already have small numbers. Or, perhaps, like the company Diversity Inc., *AncestryDNA is* building new subgroups within their algorithmic structures.[48] To address this "miscellaneous" question, *AncestryDNA* states that if there is not "sufficient data" for a particular region, then your ethnicity estimate will reflect the neighboring region(s).[49] Moreover, some populations do not differ enough on a genetic level to create different regions, which is further complicated by the fact of national boundaries changing over time. *23andMe* has an "Unassigned" category for those pieces of DNA that "match many different populations from around the world" or that "do not match any reference population very well."[50] Which brings me to Benjamin's question, "Which humans are prioritized in the process?"[51]

Many genetic testing sites include information explaining that a user's results are an estimate; but, like *AncestryDNA* and *23andMe*, even those estimates change. Some are "improved" based on how many consumers purchase and submit a kit. *AncestryDNA* explains that the ethnicity estimate is "based on probability, statistics, shared DNA, and ongoing research and science," adding that they are working on "the cutting edge of science—and in a field that is evolving rapidly" to explain their updates and changes.[52] But it is important then to question what users/consumers are buying into—with both their money and DNA. Are users simply paying to become part of ongoing research and development? That seems unlikely. So we must also ask, why does the public find these "ethnicity estimates" or the assignment of "biogeographical ancestry" so appealing?

All of the sites tend to focus on and promise to tell users something about their identity: your DNA analysis will tell you who you are. Some, like *African Ancestry*, take a very serious approach in their marketing and explanation. Their website homepage begins with a *TEDx* video where co-founder Gina Paige explains that "knowing where you're from is a critical component of knowing who you are . . . and who you can become." She frames her potential clients as "the original victims of identity theft," because "history has stolen your identity."[53] Paige makes it clear that identity is crucial for those of African descent *to know who they are* in order to move forward with their lives and culture. Some services take a lighter view of the process. One of the geneticists that Letzter interviewed even said that he sees these tests as entertainment.[54]

For example, *CRI*'s website begins with a video that is meant to be less serious and more entertaining. *CRI's* spokesperson, an older white man, explains that "you don't really know who you are or where you're from" and that "you may be completely wrong about your identity."[55] He frames the *CRI* kit as a time machine and references other DNA test pie charts as less accurate than *CRI* results, which is clearly a criticism of *AncestryDNA*. He explains that "1300 years of genealogy doesn't cut it" and actual geneticists work with your information. Besides that, *CRI* has a great Facebook score. Is this the Disneyfication of the genetic testing industry?[56] Moreover, your *CRI* results will show "how you will be known." Both *African Ancestry* and *CRI* boast of geneticists on the team, process their own lab results, and do their own data analysis; but they frame the idea that this science will show you your identity differently.

There are, therefore, both serious and entertaining aspects of the appeals to establish identity in this way. While companies sell DNA testing kits as a purveyor of individual identity the results are confounded by the movement and mixing of people over time and are never more than an estimate, especially on an individual level. One may seek out a cheap and easy understanding or confirmation of identity, but what of the consequences when what you find out is not what you are looking for? When discussing our lives online, Morville writes that "what we find changes who we become" and this seems to ring true for genetic testing.[57] People often look for confirmation of who they are through social media, family assurance, and so on. What happens when the results from genetic testing differ? Like Yiaueki and my brother, this was the case for Jesmyn Ward.

In her *New Yorker* essay about how a technology changed her life, Ward discusses her experience with *23andMe*.[58] Ward recognizes that it is "impossible for most black Americans to construct full family trees" and that genealogy falls short for those with non-European ancestry. Prior to the test, Ward "always understood [her] ancestry, like that of so many others in the Gulf Coast, to be a tangle of African slaves, free men of color, French and Spanish immigrants, British colonists, Native Americans." Ward and both her parents took the "surprisingly inexpensive" *23andMe* test. The results confirmed "proof of ancestry that they had always been denied" for Ward's parents, but she found her own ancestry disconcerting and even impossible to place within herself, preferring to define herself within the culture that she was accustomed to.[59] Lanham recognizes that "information always comes charged with emotion of some kind, full of purpose. That is why we have acquired it. The only way to make it useful is to filter it."[60] While her parents' filter was a confirmation of historical identity, Ward filtered her experience through lived cultural experience.

Identity is established through understanding of cultural experience. Ward confirms at the end of her essay that her "essential self" is "a self that understands the world through the prism of being a black American," although she struggled with her *23andMe* results.[61] Similarly, Yiaueki was shocked and had his "Indianness . . . pulled out from under [him]" after he had spent much of his life "performing Indianness."[62] Each author cites childhood practices of play or understanding of race. Each sees their race legitimized over and over through historical understanding—even when being treated cruelly or differently than other children. Cultural identity is socially constructed and founded historically. What our parents told us, the "science" of genetic testing kits, tribal legalities, data analysis algorithms, and our own recognition of culture create a strange soup of identity. Benjamin reminds us "that power operates at the level of institutions and individuals—our political and mental structures—shaping citizen-subjects who prioritize efficiency over equity."[63]

In the case of Elizabeth Warren genetic testing was condemned by some American Indians because it "undermined tribal sovereignty by equating racial science with tribal affiliation and Native American identity."[64] In other words, cultural identity cannot be found through a test. Chomsky asserts that these tests are a "way to profit from reviving and modernizing antiquated ideas about biological origins of race and repackaging them" as a "21st century version of pseudoscience that once again reduces race to a matter of genetics and origins," and I am inclined to agree. The genetic testing is not a strict science in any way and as I have demonstrated, the results may be inaccurate even beyond an individual algorithm.

Mail-in DNA kits perpetuate ideas of technological expediency and access to information that we have come to embrace culturally, and the industry of genetic testing has profited from it. Benjamin asks, "How do we rethink our relationship to technology? The answer partly lies in how we think about race itself and specifically the issues of intentionality and visibility."[65] Chomsky points out again and again that racial science is nothing new and continues to undermine minorities.[66] I will add that interdisciplinary pursuits to discover one's identity cannot just come from a company. Nobody can gift you your identity. Ironically, my brother received his DNA testing kit as a gift. I have never taken one of these tests. I know that I may have a different genetic makeup than my brother, but I also recognize that these tests cannot tell me who I am, where I am from, or who I may become—despite ample marketing.

I agree with Yiaueki and his reading of the French philosopher, Paul Ricoeur, that there is a "narrative identity" that moves us "past the question of who we are as objects and towards the question of who we are as agents."[67] Genetic testing breaks down human matter into measurable data and realigns those findings with metadata composed through various means to create different ideas about group identifiers. We all want to belong. Native Americans have fought to have agency through their tribes. Individuals, like Ward and Yiaueki, have worked through their families' understanding of cultural identity and genetic testing to become agents of their own identity, even writing their narratives and authoring their own identities. Identity cannot be found in a box or by spitting in a tube. Identity always has a history and history has always come from stories being passed down through generations whether genetically identified as true or not.

NOTES

1. Aviva Chomsky, "DNA Tests Make Native Americans Strangers in Their Own Land: Reviving Race Science Plays into Centuries of Oppression." *The Nation*, 2018. https://www.thenation.com/article/archive/dna-tests-elizabeth-warren-native-american-race-science/."

2. Ella Nilsen, "New Evidence Has Emerged Elizabeth Warren Claimed American Indian Heritage in 1986." *Vox*, 2019. https://www.vox.com/2018/10/16/17983250/elizabeth-warren-bar-application-american-indian-dna

3. Rafi Letzter. "How Do DNA Ancestry Tests Really Work?" *Live Science*, 2018. https://www.livescience.com/62690-how-dna-ancestry-23andme-tests-work.html.

4. CRI, "Frequently Asked Questions." Retrieved from: https://www.crigenetics.com

5. Chomsky, "DNA Test Make Native Americans Strangers in Their Own Land."

6. Chomsky, "DNA Test Make Native Americans Strangers in Their Own Land."

7. AncestryDNA, "Indigenous Americas Region." Retrieved from: https://support.ancenstry.com/s/article/Native-American-DNA

8. Chomsky, "DNA Test Make Native Americans Strangers in Their Own Land."

9. Ruha Benjamin, *Race after Technology: Abolitionist Tools for the New Jim Code*. Medford: Polity Press, 2019, 144.

10. Safiya Umoja Noble, *Algorithms of Oppression: How Search Engines Reinforce Racism*. New York: New York University Press, 2018,13.

11. Sequoya Yiaueki, "I Was Raised as a Native American. Then a DNA Test Rocked My Identity." *The Guardian,* 2018. https://www.theguardian.com/commentisfree/2018/nov/15/raised-native-american-dna-test-father-lied-heritage.

12. Letzter, "How Do DNA Ancestry Tests Really Work?"

13. U.S. Food & Drug Administration, "Direct-to-Consumer Tests."

14. Steven B. Katz, "The Ethics of Expediency: Classical Rhetoric, Technology, and the Holocaust." *College English*, vol. 54, no. 3 (1992), 255–275.

15. Benjamin, *Race against Technology*, 13.

16. Yiaueki, "I Was Raised as a Native American. Then a DNA Test Rocked My Identity."

17. Jeff Jaffe, "Best DNA Test in 2020: 23andMe, AncestryDNA tested and more compared." *Cnet*, 2020. https://www.cnet.com/health/best-dna-test-in-2020-23andme-ancestrydna-tested-and-more-compared/."

18. CRI, "Frequently Asked Questions."

19. Richard A. Lanham, *The Economics of Attention*. Chicago: University of Chicago Press, 2006, 19.

20. Peter Morville, *Ambient Findability*. Sebastopol: O'Reilly Media, 2005, 8.

21. Rafi Letzter, "I Took 9 Different Commercial DNA Tests and Got 6 Different Results." *Live Science*, 2018. https://www.livescience.com/63997-dna-ancestry-test-results-explained.html.

22. Letzter, Rafi. "How Do DNA Ancestry Tests Really Work?" *Live Science*, 2018. https://www.livescience.com/62690-how-dna-ancestry-23andme-tests-work.html

23. Ancestry, "More Than a Pie Chart and a Number*: Reading Your Ethnicity Estimate*." AncestryDNA Learning Hub. Accessed June 8, 2020. https://www.ancestry.com/lp/ethnicity-estimate/reading-your-ethnicity-estimate.

24. Chomsky, "DNA Test Make Native Americans Strangers in Their Own Land."

25. Letzter, "How Do DNA Ancestry Tests Really Work?"

26. Chomsky, "DNA Test Make Native Americans Strangers in Their Own Land."

27. Letzter, "I Took 9 Different Commercial DNA Tests and Got 6 Different Results."

28. Chomsky, "DNA Test Make Native Americans Strangers in Their Own Land."

29. Lanham, The Economics of Attention, 16.

30. Chomsky, "DNA Test Make Native Americans Strangers in Their Own Land."
31. 23andMe, "Reference Populations & Regions." *Customer Card.* Accessed July 5, 2020. https://customercare.23andme.com/hc/en-us/articles/212169298#h_6ab6b676-67e9-47bf-8fb8-7afea647d4f6
32. Morville, *Ambient Findability*, 125.
33. Morville, *Ambient Findability*, 152.
34. Molly McLaughlin, "National Geographic Genographic Project Review." *PC,* 2018. https://www.pcmag.com/reviews/national-geographic-genographic-project
35. National Geographic has ended the Genos project and is no longer selling kits or accepting samples for processing.
36. Nicole Smith and Rebecca Armstrong, "Best DNA Testing Kits 2020: Unravel Your Ancestry." *10 Top Ten Reviews,* 2020. https://www.toptenreviews.com/best-dna-testing-kits.
37. African Ancestry, "Home." Accessed June 17, 2020. https://africanancestry.com/home/
38. Chomsky, "DNA Test Make Native Americans Strangers in Their Own Land."
39. Benjamin, *Race after Technology*, 34.
40. CRI Genetics, "DNA Testing." *Homepage* accessed May 14, 2020. https://www.crigenetics.com/
41. Benjamin, *Race after Technology*, 30.
42. Benjamin, *Race after Technology*, 30.
43. Benjamin, *Race after Technology*, 30.
44. Letzter, "I Took 9 Different Commercial DNA Tests and Got 6 Different Results."
45. Letzter, "I Took 9 Different Commercial DNA Tests and Got 6 Different Results."
46. Morville, *Ambient Findability*, 55.
47. Ancestry, "Our Most Precise Breakdown Yet."
48. Benjamin, *Race after Technology*, 144.
49. Ancestry, "Our Most Precise Breakdown Yet."
50. "23andMe Reference Populations & Regions."
51. Benjamin, *Race after Technology*, 174.
52. Ancestry, "Our Most Precise Breakdown Yet."
53. African Ancestry, "Home."
54. Letzter, "I Took 9 Different Commercial DNA Tests and Got 6 Different Results."
55. CRI Genetics, "DNA Testing."
56. Chomsky, "DNA Test Make Native Americans Strangers in Their Own Land."
57. Morville, *Ambient Findability*.
58. Jesmyn Ward, "Cracking the Code." *The New Yorker*, 2015. https://www.newyorker.com/tech/annals-of-technology/innovation-cracking-the-dna-code
59. Ward, "Cracking the Code."
60. Lanham, *The Economics of Attention*, 19.
61. Ward, "Cracking the Code."

62. Yiaueki, "I Was Raised as a Native American. Then a DNA Test Rocked My Identity."

63. Benjamin, *Race after Technology*, 30–31.

64. Adam Elmahrek, "Elizabeth Warren Again Is Pressed on Past Claims of Native American Heritage." *Los Angeles Times*, 2020. https://www.latimes.com/politics/story/2020-02-26/elizabeth-warren-again-is-pressed-on-past-claims-of-native-american-heritage

65. Benjamin, *Race after Technology*, 54.

66. Chomsky, "DNA Test Make Native Americans Strangers in Their Own Land."

67. Yiaueki, "I Was Raised as a Native American. Then a DNA Test Rocked My Identity."

BIBLIOGRAPHY

23andMe. "Reference Populations & Regions." *Customer Card.* Accessed July 5, 2020. https://customercare.23andme.com/hc/en-us/articles/212169298#h_6ab6b676-67e9-47bf-8fb8-7afea647d4f6

African Ancestry. "Home." Accessed June 17, 2020. https://africanancestry.com/home/

Ancestry. "Indigenous Americas Region." *Ancestry Support.* Accessed July 5, 2020. https://support.ancestry.com/s/article/Native-American-DNA

Ancestry. *"More Than a Pie Chart and a Number: Reading Your Ethnicity Estimate."* AncestryDNA Learning Hub. Accessed June 8, 2020. https://www.ancestry.com/lp/ethnicity-estimate/reading-your-ethnicity-estimate

Ancestry. "Our Most Precise Breakdown Yet." *DNA Ethnicity Estimate Update.* Accessed June 11, 2020. https://www.ancestry.com/dna/lp/ancestry-dna-ethnicity-estimate-update

Benjamin, Ruha. *Race after Technology: Abolitionist Tools for the New Jim Code.* Medford: Polity Press, 2019.

Chomsky, Aviva. "DNA Tests Make Native Americans Strangers in Their Own Land: Reviving Race Science Plays into Centuries of Oppression." *The Nation,* 2018. https://www.thenation.com/article/archive/dna-tests-elizabeth-warren-native-american-race-science/

CRI Genetics. "DNA Testing." *Homepage* accessed May 14, 2020. https://www.crigenetics.com/

Elmahrek, Adam. "Elizabeth Warren again is pressed on past claims of Native American heritage." *Los Angeles Times*, 2020. https://www.latimes.com/politics/story/2020-02-26/elizabeth-warren-again-is-pressed-on-past-claims-of-native-american-heritage

Jaffe, Jeff. "Best DNA Test in 2020: 23andMe, AncestryDNA tested and more compared." *Cnet*, 2020. https://www.cnet.com/health/best-dna-test-in-2020-23andme-ancestrydna-tested-and-more-compared/

Katz, Steven B. "The Ethics of Expediency: Classical Rhetoric, Technology, and the Holocaust." *College English*, vol. 54, no. 3 (1992), 255–275. JSTOR.

Lanham, Richard A. *The Economics of Attention.* Chicago: University of Chicago Press, 2006.

Letzter, Rafi. "I Took 9 Different Commercial DNA Tests and Got 6 Different Results." *Live Science*, 2018. https://www.livescience.com/63997-dna-ancestry-test-results-explained.html

Letzter, Rafi. "How Do DNA Ancestry Tests Really Work?" *Live Science*, 2018. https://www.livescience.com/62690-how-dna-ancestry-23andme-tests-work.html

McDermott, Marc. "Best DNA Test Kits." *Smarterhobby*, May 11, 2020. https://www.smarterhobby.com/genealogy/best-dna-test/

McLaughlin, Molly. "National Geographic Genographic Project Review." *PC,* 2018. https://www.pcmag.com/reviews/national-geographic-genographic-project

Morville, Peter. *Ambient Findability.* Sebastopol: O'Reilly Media, 2005.

National Geographic. "FAQ: End of Kit Sales and Previously Purchased Kits." *Geno 2.0.* Accessed June 13, 2020. https://genographic.nationalgeographic.com/faq/sales-shutdown-previous-kits/

Nilsen, Ella. "New Evidence has Emerged Elizabeth Warren Claimed American Indian Heritage in 1986." *Vox*, 2019. https://www.vox.com/2018/10/16/17983250/elizabeth-warren-bar-application-american-indian-dna

Noble, Safiya Umoja. *Algorithms of Oppression: How Search Engines Reinforce Racism.* New York: New York University Press, 2018.

Smith, Nicole and Rebecca Armstrong. "Best DNA Testing Kits 2020: Unravel Your Ancestry." *10 Top Ten Reviews,* 2020. https://www.toptenreviews.com/best-dna-testing-kits

U.S. Food & Drug Administration. "Direct-to-Consumer Tests." 2019. https://www.fda.gov/medical-devices/vitro-diagnostics/direct-consumer-tests

Ward, Jesmyn. "Cracking the Code." *The New Yorker*, 2015. https://www.newyorker.com/tech/annals-of-technology/innovation-cracking-the-dna-code

Yiaueki, Sequoya. "I Was Raised as a Native American. Then a DNA Test Rocked My Identity." *The Guardian,* 2018. https://www.theguardian.com/commentisfree/2018/nov/15/raised-native-american-dna-test-father-lied-heritage

Chapter 8

Technologies of Convenience

An Examination of the Algorithmic Bias in the Input/Output System of Digital Cameras

Joel S. Beatty

Society faces a milieu of changing technologies that prompts serious questions about digital cameras used for purposes of surveillance of its citizens. We have been warned of the inherent dangers of technological artifacts such as digital images, artificial intelligence, facial recognition, big data, and biased algorithms. Humanities scholars have warned us.[1] Scientists and engineers have warned us.[2] Human rights organizations have warned us.[3] Lawyers have warned us.[4] Digital cameras are being used as means of technological control to reduce individual human beings into bits of computational data. We've also been warned that a range of biases can permeate these systems of technology, many of which are hidden or beyond user control. The social effect of this work invites cultures to wonder about a future of living in this technological milieu where digital cameras are ubiquitous and digital imagery is central to the inchoate effects of machine learning, artificial intelligence, and technical systems built on "big data." Some of the central questions that help synthesize this particular kind of scholarship are: How has the digital camera become such a central symbol of surveillance? How have massive databases of digital images become a "focus-problem" to investigate machine learning bias, gender bias, racial, and ethnic bias? How have these systems of digital imagery gained so much sociotechnical control? More importantly, how do we untangle ourselves from the more pernicious uses of human-to-digital data?

The broad set of ideas articulated in this volume, in part, have helped frame some of the more problematic questions we have about technology by venturing into new scholarship on algorithmic culture. As Striphas explains, the scholarship of algorithmic culture examines why "human beings have been delegating the work of culture—the sorting, classifying, and hierarchizing of people, places, objects, and ideas—increasingly to computational processes" and he argues that we should focus on a widening divide between "public" participation and "exclusive and private" interests in the design of algorithms that organize cultural data.[5] In this chapter, I adopt this definition specifically to premise the development of the digital camera not just as a technology for power and control of culture in the pursuit of data, but also as a "technology of convenience" that has grasped the common consumer and ushered in new and undefined consequences for our bodies.[6]

My purpose is to interrogate the system of "bias" that enters into the algorithmic decision-making processes in the production of digital color imagery and, more specifically, images that represent human bodies. Whether it is for the purposes of art, journalism, nostalgia, surveillance, biometrics, or even capitalism, the digital camera "user" relies on an "algorithmic chain" of an input/output system where potential bias can be introduced at many different points.[7] Critical understanding of where bias occurs along this chain is of a high value for an algorithmic culture that includes engineers, policy makers, technical communicators, and a consuming public alike.[8] Toward this end, this chapter investigates the processes of designing algorithms for the digital color input/output system and will pose critical questions about the "algorithmic logic" that contributes to the emergent cultural and subjective landscape of digitally colored imagery. Ideally, this perspective can help inform critical consumers of digital cameras and offer lines of inquiry into the higher order issues of technological surveillance and control. Yet, I argue that understanding the algorithmic culture has built this technology on the value of convenience is the first step toward that inquiry.

BIAS AND ALGORITHMIC CHAIN

What is an algorithmic bias? How can it be differentiated from other types of bias? Whether algorithmic systems are autonomous or semi-autonomous, or if humans are fully "in the loop" in the decision making of an algorithmic process, the potential for multiple biases along the full chain of the algorithmic process exists.[9] As Danks and London explain, algorithm designers use a taxonomy of "unobjectionable" and "problematic" biases to determine if there is a need to respond to a recognized bias in a system, and in some cases "there are technological or algorithmic adjustments that developers can use to compensate for problematic bias."[10] These "compensations," however, become more complicated and difficult to detect in complex systems where it is increasingly difficult discern how the algorithm arrived at a decision. One of the central challenges to mitigating problematic biases in algorithms is that there's a "perquisite understanding of the relationship between the autonomous system and the ethical and legal norms in force in the relevant contexts."[11] To further describe algorithmic biases within a process of development, implementation and application, Danks and London categorize the biases along this chain as (1) training data bias, (2) algorithmic focus bias, (3) algorithmic processing bias, (4) transfer context bias, and (5) interpretation bias. This system allows designers to determine where a problematic bias requires mitigation along an algorithmic chain and to identify the type of adjustment that needs to be made. If there is a technical or unavoidable limitation to eliminating the bias, the designers can "balance" or "compensate" for the bias by making adjustments in another part of the system.[12]

This work has been useful for scholars investigating the presence of gender, racial, and ethnic bias in software systems that rely on algorithms and data. Silva and Kenney have recently put forth a model they call the Algorithmic Value Chain, which can help identify where a potential bias can be introduced. Since preexisting human biases coded within an algorithmic system are not mutually exclusive, they contend that isolated biases interact within a singular algorithmic process of Input > Algorithm > Output > Users > User Modification/Feedback.[13] Recognized biases, then, can emerge within each "node" of the chain. The "algorithmic" biases Silva and Kenney describe are:

1. Input
   a. Training Data Bias
   b. Algorithmic Focus Bias
2. Algorithm
   a. Algorithmic Processing Bias
3. Output
   a. Transfer Context Bias
   b. Interpretation Bias
   c. Outcome Non-Transparency Bias
4. Users
   a. Automation Bias
   b. Consumer Bias
5. User-Modified Data
   a. Feedback Loop Bias

The authors proposed this visual model as a way for computer scientists and users alike "to address the serious social problem of bias . . . and for identifying where bias might emerge in a complex interaction between algorithms and humans." The value of this model, they argue, is that it offers a way to understand how the social impact of biases are expressed and reinforced by digital technology.[14] It is important to note that the Algorithmic Value Chain model (AVC) is not intended to eliminate algorithmic bias, but is a helpful tool to detect where gender, racial, ethnic, and other forms of discriminatory biases interact with algorithmic data systems. In the analog world, discriminatory biases are fluid and sometimes difficult to detect and track, but when discrimination is coded into algorithms and data sets, the very same algorithms and data "create a record" and hold the potential for detection and analysis of these biases.[15]

While the ACV model is valuable to engineers and policy makers, what about the common consumer that lacks any knowledge of known bias in an algorithmic system? Without transparency and literacy in the algorithmic logic of a complex system, it is of little wonder that sources of biases are conflated by users without any agency in the system. I argue that the AVC model offers an entry point for the common consumer of digital cameras to make informed decisions about data captured in digital images by providing a roadmap for how biases interact in the digital process of image making.

ALGORITHMIC BIAS AND THE INPUT-OUTPUT SYSTEM OF DIGITAL IMAGES

Admittedly, my brief analysis in this chapter cannot fully address the ACV model, nor can it test the model in a way that validates or invalidates it. However, in the following section, I focus on some applicable knowledge and historical synthesis of digital cameras that is relevant to the everyday user and can open lines of inquiry into algorithmic bias and identify where problems can occur along the digital image process. To be more precise, I focus on algorithms and digital processes that are used to process images of the human body (i.e., skin, hair, eyes). I also review just one of the many problems that engineers and developers address in the application of an algorithm that every digital camera relies on: the "color space" algorithm.

To create a digital image, several general and technological phases of the digital camera's process must be understood.[16] First, a digital camera technology must be used for *sampling* an analog scene, and, second, therefore have a mechanism for capturing light and converting light information into digital information, also known as *encoding*. A third phase is the *transmission* of that code, or signal, to a computing device, which then requires an additional transmission to a display device. To then see an image on a screen, the code/signal must go through a *reconstruction* phase that will convert the code back into light intensities that we see on a display screen. This process is also known as the *input/output* process and can be visualized in the following scheme (see figure 8.1).[17]

Signal processing is the term used for the algorithmic conversion of digital data after the input of a sample, and this is where color matching technology

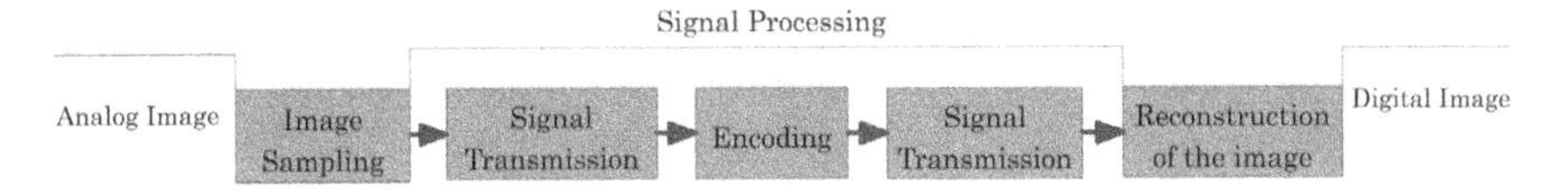

Figure 8.1 The Linear Input/Output System—from Analog to Digital Imaging. Diagram courtesy of the author.

becomes important to the process. The color system of a sampling device (a camera) may be different than the color system used in the reconstruction of an image (a digital screen). In order to ensure that colored images will match between sampling and reconstruction, the CIE color space provides a numerical grid to allow the color values of the input system to be converted to the color values of the output system in a way that, in theory, will maintain perceptive continuity.

The technical devices needed to facilitate the input/output digital image process are somewhat fluid. For instance, a digital camera is a sampling device that initially processes information then transmits it to a computer for additional processing and encoding with a graphic card. However, some cameras have both input and output capabilities, which sample, encode, and directly transmit a signal to a specifically designed screen that completes the image reconstruction process. Color management is an important part of this process because color processing remains independent to each device in the system. For this reason, all technologies in the input/output image making system must be compatible for the system to work. Because the structure of analog to digital, input to output imaging system relies on successful interface and translation of light information between technologies, and because the user interface between the screen and the perceiver demands an optimum reconstruction of an analog image, the system of color management from sampling to reconstruction has been identified as the Color Appearance Model (CAM).[18]

Fairchild describes a CAM model as containing both linear and nonlinear processes, and he defines it as a "model" due to (1) the complexity of its parts, which include psychophysical, biophysical, and technological interactions, and (2) a user of the model can make a prediction about how analog images will appear post-reconstruction.[19] The main premise behind this technical theory is that a system user (or automated technology) makes a prediction of how a reproduced color (or colored image) will appear to a human observer. A color technology using an automated CAM process can produce an image that either (1) visually matches an analog sample, or (2) visually matches the intent of the designer. Most importantly, a designer using a CAM, once mastered, can reliably produce images that maintain color continuity for a broad range of user. However, to maximize perceptive continuity for the broadest range of users possible, color systems using an algorithmic CAM were built for what is referred to as a "standard observer" based on what is considered "normal" trichromatic color vision.

Colorimetry refers to the measurement of color appearance for human observers, which is based on the measurement technology of spectrometry that measures light at three different electromagnetic wavelengths (also known as *trichromacy*). In other words, the practice of colorimetry measures the "human" visible spectrum of light in units of distinguishable wavelengths in order to predict the way our visual system interprets light.[20] The purpose of colorimetric matching technology is to make colors that are displayed in different media, and viewed under different lighting conditions, visually coherent.

One of the biggest advancements in the practice of colorimetry that led to standardized CAMs stemmed from a series of reports by the Optical Society of America (OSA) in 1922 and the *Commission Internationale de L'éclairage* (CIE) in 1931.[21,22,23] The OSA and CIE reports relied on the latest data from vision science as a way to create a "profile" for what they called "the standard observer," which included their stated goals of developing (1) a standard color terminology, (2) a standard interpretation of psychophysical data in relation to light stimulus, (3) a standard for measuring radiance, or light intensity, (4) a standard for the methods of measurement, and (5) a standard of comparison between these different scales.[24] The early work from OSA's report in 1922 on colorimetry led to separate experiments in the late 1920s by David Wright and John Guild, who devised a measuring device and method that could observe tristimulus responses of the human retina by placing the device close to the fovea.[25] They carefully measured a very small number of human subjects based on the spectral primaries established by James Clerk Maxwell: their stated goal was to accurately define the spectral sensitivities of visual receptors.[26] With the "standard observer" already defined in the 1922 OSA report on colorimetry, Wright and Guild's data enabled the creations of the CIE and XYZ color algorithms that could be used to "visually match" RGB trichromatic values represented by the "standard observer."[27,28] This data led to the creation and adoption of the CIE's color system and a functionally adaptable "Chromaticity Diagram" at their 1931 meeting in Cambridge, England.[29]

The "standard observer" has not been without controversy among practitioners of optics throughout the history of the CIE, leading to several revisions of the criteria to improve its accuracy.[30] Most notably, major epistemological shifts in the theories of human colored vision from other areas of science have occurred since 1931 and significantly contributed to the way the CIE's color matching technology is understood to work. The first CIE color chromaticity chart was an algorithmic model representing the visual spectrum that a human with normal trichromatic color vision can see.[31] This model broke down the visual spectrum into data points on a X,Y,Z axis. The science of colorimetry, then, allowed for a sample of any part of an analog scene to be assigned a tristimulus value, usually expressed in red, green, and blue (RGB) values. The RGB values also putatively correspond to the light-sensitive cones in the normal human eye (long, medium, and short wavelengths, respectively) and can be accurately converted to CIE color space by converting RGB to XYZ conversion values. Therefore, in theory, any reconstructed color using RGB values for its output will visually match the input sample.[32]

The problem engineers of the CIE and OSA faced in using trichromatic RGB values was realized when mapping the values onto a graph, where both positive and negative values are shown along with the R (long) wavelength.[33] This means that spectral reflectance measurements and the RGB tristimulus values, alone, could not account for the range colors that typical humans can see, and this was a problem those early engineers and mathematicians had to overcome algorithmically. The collected reflectance measurements alone could not account for the full spectrum of "color" as it appears to the standard observer, instead, an algorithm projected a universal vision and an expansive color space.[34]

It should be noted that algorithmic conversion of RGB values to XYZ color space enabled consistent visual matching and more accurate interactions between systems that assured more accurate output. While RGB color is based on physical measurements of tristimulus wavelengths, the XYZ color space was purely a theoretical model.[35] A mathematical theory was needed to express the tristimulus values in positive terms that would show up on the spectrum. One big achievement of the CIE report in 1931 was figuring out the math to convert tristimulus values based on data from human visual testing into XYZ terms that could be mapped on a 2-D vertical plane.

Essentially, these mathematical algorithms allowed analog RGB values to be viewed in three-dimensional (3D) space and to be broken into three separate values of hue, saturation (chroma), and brilliance (aka value, intensity, or luminance). The creation of a 2D plane (visualized on a graph) was helpful for users working with color because it separates hue and saturation from the third axis of light intensity, which is on a scale from light to dark.[36] Neurobiologist, Margaret Livingstone, notes that all 2D representations of color, including Newton's color wheel, share the same feature by excluding luminance.[37] So, what is shown in the 2D CIE space is of functional importance for understanding the algorithm. CIE color space standardized the color spectrum by giving it a numerical value, which also worked well with existing imaging technology already designed for the grayscale of light to dark (luminance). The underlying logical importance for CIE's color algorithm, then, is in symbolic operations that made the reconstruction calculations reliable, thus knowable. The CIE algorithm, then, made the task of replicating colors more consistent throughout many systems that rely on producing colors.

The 2D representation of color space also addressed another technical limitation, which was that RGB values expressed in this way did not account for all the mixtures of hues between the long (R) and the short (L) wavelengths the standard observer can see. To state this in plain color language: tristimulus values do not account for all the mixtures of blues and reds, including the pinks, many of the purples and violets, that people with "normal" trichromatic color vision can perceive. The RGB to XYZ conversion, however, was able to account for this because, like Newton's color wheel, it joined the visual spectrum together in both 3D and 2D Color Space.

To explain the physiological importance of this color model, consider an image of a pinkish or "rosy" human skin. The tristimulus (RGB) system of measurement calculates RGB values into single points of the light wavelengths, but since there is no single spectral wavelength that appears to the standard observer as "pink," it can be inferred that pink must be a mixture of wavelengths. The CIE color system allowed a more accurate method to calculate what mixture of two spectral colors (red and blue) it takes to produce a perceived non-spectral color (pink). The CIE's calculations for RGB to XYZ conversion become critical to digital imagery in the *encoding* phase because in the *reconstruction* phase digital images are once again given RGB values containing the image information we see on our display screen. If our vision was only a matter of three primary tristimulus values, those with normal trichromatic vision wouldn't be able to perceive many of the mixtures of red and blue light that make up our visible spectrum. Similar technical limitations of this system exist in the algorithmic processing of darker skin. Since the problem of light luminance (light/dark intensity) had to be addressed algorithmically, the conversion of darker skin tones is less accurate than lighter skin tones due to a number of physical and technical limitations.

The developers of the CIE algorithm realized there was still something in the way that humans perceive color that must be accounted for, which required them to solve a technical problem in the original OSA and CIE reports on color in order to make and functionally predictive model.[38] The algorithmic "problem" with the CIE color space can be described from a current perspective as a technical constraint coming under a biophysical constraint, and, based on the language used to describe the problem, this is something the writers of 1922 *Report on Colorimetry* were aware of. The report states:

> It is impossible to identify color with radiant energy, or wavelengths of radiant energy, although energy is the adequate stimulus for color. This is because color is known to depend upon the presence of the perceiving individual and

> because it is directly recognized to be something radically different in kind from its stimuli. Consequently, nothing but confusion can result from the use of the word "color" as a synonym or wavelength. . . . Color cannot be identified with or reduced to terms of any purely physical conception; it is fundamentally a psychological category.[39]

Here the language infers a philosophical approach to vision that suggested a strict division between subject and object; and consequently, the problem rested on a fundamental divide between psychological and physical processes. In terms of algorithmic logic, the authors of the report identified a new set of postulates that limited their definition "color" to a psychological phenomenon, but operationally they were able to solve this technical dilemma in a way that met criteria for logical coherence as applied technology.

After an update to CIE's criteria for the "standard observer" in 1964, the criteria were updated again 1976 and developed in two models known as the CIELAB and CIELUV color models that included a synthesis of terminology advanced from opponent color theory.[40] The updated CIELAB color space of 1976 was also compatible with the XYZ and RGB values derived in 1931; thus, the CIE adopted the CIELAB color space as a uniform model based on an revised data for a standard observer.[41] "L" stands for luminance, "A" represents the red green/axis, and "B" represents the yellow/blue axis. The development of this space, then, synthesized the trichromatic and color opponent understandings of color vision. The adaption of the CIELAB space allowed color differences "to be perceptually uniform throughout the space."[42] The CIELAB color space is still used as an effective algorithm to this day. However, different color models with visual matching functions—designed for specific purposes and industrial uses—have been created using the same basic methods. The development of the XYZ dimensional space that allowed color to be mapped on a two-dimensional plane is the basis for all color spaces used in CAMs. Additionally, if you have the 3D values of any color space, analog or digital, the algorithmic XYZ space can be used to convert data into any other color space. When the first digital cameras were developed in the 1960s and 1970s, the CIE color space was found to be an adaptable technology that could be easily integrated into the digital image-making process. All of this algorithmic computing now takes place in the computer with minimal input from the user, but this is how visual data is transferred between most input/output systems today.[43]

This complex history is of critical importance for understanding algorithmic bias in the digital image processes, because it traces the just one of the many "standards" by which algorithmic bias is measured. However, using the ACV model, determining whether the bias is "unobjectionable" or "problematic," thus requiring a human response, depends on the use and implementation of the algorithmic technology. While most uses of digital cameras are innocuous, the same digital process that produces a beautiful portrait photograph of a human being can also enable the computational data of that photograph to be expressed as a recognizable bias anywhere along the algorithmic chain.

BIAS AND THE DIGITIZING OF THE HUMAN BODY

Most smartphone users don't think of engineering standards, algorithmic chains, data biases, or their camera as a "prediction model" when they use it to snap a photograph, but according to a growing body of literature on algorithmic bias, the widespread use of digital image processing from cell phone cameras has prompted an explosion in data sampling, collection, and implementation of algorithms built to digitize information captured from the human body.[44] The CIE color space and its progeny of predictive computational algorithms allow the human body to be enumerated, digitized, classified, and collected into data sets. For example, a software engineer can write an algorithm to detect all the pixels in an image that represent skin color.[45] Because each pixel of a digital image can be algorithmically assigned an RGB color value, that number can be mapped to corresponding color on the CIE color space.

For the purpose of this scenario, imagine taking a "selfie" with an *iPhone 11*. It creates an image with a resolution of 2316 × 3088; that image would have over seven million individual pixels, each with an assigned RGB number. For most images, this data is algorithmically processed for an output of a digitized image. Along this typical algorithmic chain, engineers and developers have addressed many non-objectional biases to compensate for the technical limitations. Maybe they have added color correction controls that allow the user to manipulate the skin tones in the selfie. Perhaps the camera has a "skin tone" or "white balance" setting that adjusts for a range of skin colors somewhere in the algorithmic process. In this context, these are non-objectional, or neutral algorithmic biases, with a mixture of autonomous and semi-autonomous controls over the system. But there's an alternative path for this data.

Now imagine that your "selfie" is placed into a facial recognition system with a Human Computer Interface (HCI). Most HCIs are designed with algorithms to extract the pixels of the image that represent a range of skin colors and their correlating numbers in the color space.[46] Bias is present in each node of this algorithmic chain, because in order to develop a process that detects skin color, developers need to account for many machine biases to achieve the most accurate representation. Most of these systems compensate for variables of bias that stem from physical and technical issues like scene illumination, image device characteristics, and the background or nonhuman objects in the scene; however, developers that work with this technology also train algorithms to detect human variables like ethnicity, age, gender, hairstyle, makeup, and skin ailments.[47] Facial recognition features built into smartphones allow developers to build AI systems that rely on machine learning algorithms where the data are already skewed with gender and racial bias. Buolamwini and Gebru explain that face detection platforms are not only used to code gender but also to classify "phenotypes" based on skin color, and this is a problem because most research shows accuracy rates in the detection of female faces and black faces are much lower than in their counterparts.

In applying an ACV model analysis, this problem can be described as a "data training bias" because facial recognition platforms use massive databases of facial images coded with demographic information in order to match to an image taken with facial recognition software. In this scenario, the skin colors in your "selfie" are extracted, assigned a numerical profile, and matched to other images in the database using a predictive algorithm. In most cases, big commercial platforms using facial recognition also store your data and incorporate it into a feedback loop to improve the efficacy of their databases and automated algorithms, but due to the competitive commercial market to develop better facial recognition apps, private companies do not publish all of their metrics and methodologies. The result makes the algorithmic bias opaque to the end-user. Yet even with this lack of transparency, quality control researchers (like Buolamwini and Gebru) can test the technologies for errors, determine where bias exists, and suggest areas of improvement in the algorithmic chain. Nonetheless, the average consumer of digital cameras and smartphones is not "in the loop" of information and the decision-making processes over the data being collected. This lack of transparency is reinforced by the division between public and private interests.

So how does a problematic racial bias enter into a seemingly non-objectional system of algorithmic biases like digital image processing? In *Race after Technology*, Benjamin boldly (and in my opinion, correctly) claims:

> Human toolmaking is not limited to the stone instruments of our early ancestors or the sleek gadgets produced by modern tech industry. Human Cultures also create symbolic devices that structure society. Race, to be sure, is one of our most powerful tools—developed over hundreds of years, varying across time and place, codified into law and refined by custom, and tragically, still considered by many people to reflect immutable differences between racial groups. . . . Race as technology: this is an invitation to consider racism in relation to other forms of domination as not just an ideology or history, but as a

> set of technologies that generate patterns of social relations, and these become black-boxed as natural, inevitable, automatic.[48]

There is a fundamental lesson in the history of the CIE color algorithm that we can obtain from this view of "race as technology": the algorithms of technical color play a huge role in structuring human and racial relationships in society, both then and now, because it allows the digital imaging process to embody preexisting racial relationships. Further, as the systems of image processing evolved to be more autonomous, the technical limitations of the digital imaging technology became codified as "natural" limitations. After all, long before the CIE algorithm existed, societies of the past had already codified "race" into shades of skin color. When a technology evolved that could enumerate these symbolic shades of human variation and represent them in a machine, it allowed technology to become a tool of racism while masking the technological limitations. The algorithm, in this view, becomes an extension of "nature"—something we succumb to—rather than "cultural"—a social problem that we in fact can mitigate.

DIGITAL IMAGE PROCESSING AND TECHNOLOGIES OF CONVENIENCE

In the beginning of this chapter, I argued that the development of digital cameras along with their system of algorithms, databases, and technical processes, was driven by the underlying value of convenience rather than power and control. On the surface, the idea that the CIE colors space is a "convenient technology" makes sense: the technology addressed a significant problem of matching industrially made pigments, dies, and paints. The "algorithmic culture" that developed it had to build a technology that was in alignment with standard human visual perception and easily adapted to different technologies. The CIE algorithm and the numerical computation of color helped solve a global industrial problem, and its convenience can be measured by its widespread applicability. As Tierney explains, convenient technologies establish order through every facet of society by mapping out and expanding the in use through widespread applications; and it is because of its commitment to technological convenience that some technologies encounter very little resistance to their power.[49] Slack and Wise extend the argument against the simplistic view that technologies are merely "built for convenience" and "built to meet the demands of the body." Instead, they argue that ever-changing meanings of convenience and comfort, these meanings correspond to new ways people relate their bodies to technology.[50] In this view, technologies are not only built to meet the demands of the body but "to overcome the limits of the body," in particular the relationship our bodies have with space in time. In essence, the CIE color space algorithm allowed "color" to become "tele-present" on a global scale through a computational formula, thus overcoming challenges of space and time while meeting the demands of the body. The algorithm has been modified since 1931, but not replaced. As digital cameras and image processors developed, the CIE color space developed into widespread applications due to its value of convenience. Now, inside every digital camera, there's an automated algorithm to digitize and enumerate analog information, with limited user control. A digital picture of a human body can become a beautiful piece of art or capture human emotions in a way words cannot, but it can also turn our bodies into digital data. Central to this argument, consequently, is an examination of how little resistance there is to the power of the digital image process. Even with its technical limitations, the CIE algorithm enabled numerical coding of any image, giving visual information and data the power of tele-presence.

The consumer should take notice of the warnings in a growing literature on the inherent bias of big databases and the algorithms built to process, sort and categorize the data. It is of vital importance for the end user of these systems to stay "in the loop" as the humans with the most at stake. Yet, the average consumer seems to be caught in a powerless position and subservient to an algorithmic culture that builds new technologies faster than the consequences can be realized. The responses to this power are justified. But it is also a myth that the common consumer of digital technology cannot play a critical role in the mediation of algorithmic bias, by understanding where bias enters the algorithmic chain in the digital image process. The historian of technology, Thomas Hughes, once wrote that "a technological system can be both a cause and an effect; it can shape and be shaped by society. As they grow larger and more complex, systems tend to be more shaping of society and less shaped by it."[51] It's between these poles of technological determinism that most consumers of digital technologies see themselves, which sets up a struggle for control. The view of technological convenience, on the other hand, can empower a digital citizen to recognize the value of slow growth and balance within a system. Being the critical consumers of digital cameras means that we have a vital role to play in a system that imports human biases into autonomous and semi-autonomous loops that our culture is responsible for.

NOTES

1. Ruha Benjamin, *Race after Technology*. Medford, MA. Polity Press (2019). Print.—Benjamin makes the case that a "default whiteness" is coded into photography and image processing. She articulates the history of commercial kodak film that

by default was set for lighter skin, and then traces this system of racial bias to the digital algorithms of Google image searches.

2. Buolamwini, Joy and Gebru, Timmit "Gender Shades: Intersectional Accuracy Disparities in Commercial Gender Classification." *Proceedings of Machine Learning Research* 81:1–15 (2018). Print.—The authors conduct an "accuracy audit" of the data sets for facial analysis algorithms. These data sets are widely used by commercial platforms like Google, IBM, Microsoft, and Apple. What they determined is a form a "data training bias" that is less accurate for "female" and "black" faces than from other image sets based on "phenotype" traits. They call for more transparency from private companies to provide their data sets and explore way to address algorithmic bias.

3. In the report "Surveillance Giants: How the Business Model of Google and Facebook Threatens Human Rights" (2019), Amnesty International presents evidence of how the business giants of Google and Facebook are exposing the vast amount of data they collect from users to possible human rights and privacy violations. The data is mostly collected from AI system that have become ubiquitous in the platforms and it amounts to an assault on privacy that creates imbalance of private and public interests in data services, including grabbing "meta-data" from all the images you store, send and receive on your connected devices.

4. Clare Garvie, Alvaro Bedoya,, and Jonathan Frankie. "The Perpetual Line-Up: Unregulated Police Face Recognition in America.". *Georgetown Law Center on Privacy and Technology.* Georgetown Law's Center on Privacy & Technology (2016). The GLCP report details the possible civil rights violations that image data collection from public and private cameras presents to our legal system.

5. Ted Striphas "Algorithmic Culture." *European Journal of Cultural Studies* 18(4–5): 395–412 (2015).

6. Jennifer Daryl Slack and MacGregor Wise. *Culture and Technology: A Primer,* 2nd edition. New York, NY: Peter Lang (2015).

7. Silva Selena, and Martin Kenney. "Algorithms, platforms, and Ethnic Bias: A Diagnostic Model." *Communications of the Association of Computing Machinery.* (2019).

8. Tarleton Gillespie, "The Relevance of Algorithms." In *Media Technologies: Essays on Communication, Materiality, and Society.* ed. Tarleton Gillespie, Pablo Boczkowski,and Kirsten Foot. Cambridge, MA: MIT Press. (2014).—Gillespie calls for a "interrogation of algorithms as a key feature in our information ecosystem." He highlights "six dimensions of public relevance" that can help us map lines of inquiry into algorithmic activity including (1) Patterns of inclusion, (2) Cycles of anticipation, (3) The elevation of relevance, (4) The promise of algorithmic objectivity, (5) Entanglement with practice, and (6) The production of calculated publics.

9. David Danks and Alex J. London, "Algorithmic Bias in Autonomous Systems," in *Proceedings of the Twenty-Sixth International Joint Conference on Artificial Intelligence* (August 2017), 4691–4697.

10. Dank and London, "Algorithmic Bias in Autonomous Systems," 4691.

11. Ibid., 4696.

12. Ibid.

13. Silva and Kenney provide a salient visualization that synthesizes their model with Danks and London's description of the Algorithmic Chain.
14. Silva and Kenney, 8.
15. Ibid.
16. Phil Green, *Understanding Digital Color*. Sewickley, PA: GAFT Press (1999).
17. See Bunge (1966). *Technology as Applied Science*, p 33—Bunge says that when researchers describe their technological system in terms of input/output through a change to the external variables of a system, this is essentially an attempt to schematize their system as a "black box" that allows them to sever the "ontological import and ignore the adjoining levels."
18. Mark D. Fairchild. *Color Appearance Models*. Hoboken, NJ: John Wiley and Sons (2005)—Fairchild is an Imaging Scientist and professor at the Rochester Institute of Technology.
19. Ibid., 183.
20. The three wavelengths are also referred to in this paper with the symbols S, M, and L which refer to the wavelengths expressed in nanometers: S = 420 nm–440 nm, M= 530 nm–540 nm, L = 560 nm–580 nm. These short, medium and long wavelengths also roughly correspond with the Red, Green, and Blue (RGB) trichromatic color space.
21. L. T. Troland, "Report of Committee on Colorimetry." *Journal of the Optical Society of America and Review of Scientific Instruments* 6(6) (1922), 527–596.
22. Fairchild, *Color Appearance Models*," 53–82.
23. The CIE translates in English to International Commission on Illumination which is sometimes referred to in the literature as ICI.
24. Troland, "Report of Committee on Colorimetry," 530.
25. W. D. Wright, "A Trichromatic Colorimeter with Spectral Primaries." *Transactions of the Optical Society* 29(5), 225–242 (1928); W. D. Wright. "A Re-Determination of the Trichromatic Coefficients of the Spectral Colours." *Transaction of the Optical Society* 30(4): 141–164 (1929); J. Guild. "The Colorimetric Properties of the Spectrum." *Philosophical Transactions of the Royal Society of London* 230: 149–187 (1932).
26. Lee (2008), p. 8.
27. Loyd A. Jones, "The Historical Background and Evolution of the Colorimetry Report." *Journal of the Optical Society of America A* 33(10): 545 (1943).
28. Fairchild, *Color Appearance Models*, 76.—Fairchild notes that the original 1931 color space was based on the measurement of fewer than twenty human subjects.
29. "History of the CIE 1913–1988" Photocopy Edition. Vienna, Austria: Comission Internationale de L'eclairage, CIE Central Bureau. Web. Retrieved on 7/22/2017. WWW: http://www.cie.co.at/ (1999), 14.
30. For two in-depth critical reviews of the CIE's original color system in 1931, see Broadbent (2003). "A Critical Review of the Development of the CIE1931 RGB Color Matching Functions" and Fairman, H. S., M. H. Brill, and H. Hemmendinger (1997). "How the CIE1931 Colormatching Functions Were Derived from the Wright-Guild Data."
31. "History of the CIE 1913–1988" (1999).

32. Maxwell, James Clerk, "Colour Vision." *Scientific Papers,* Vol. 2, Edited by W. D. Niven Cambridge, England: Cambridge University Press, pp. 267–285 (1890).—Today, color values in digital photographic technology are usually expressed in a trichromatic system known as additive color that assign each pixel a numerical value between 0 and 255 to the three separate color channels of red, green and blue (RGB). However, the RGB model can be traced back to James Clerk Maxwell's papers on color, in particular a paper title "Colour Vision" in which he describes a theory of RGB color mixing using this system.

33. Fairman, Hugh S., M. H. Brill, H. Hemmendinger, "How the CIE Color Matching Functions Were Derived from Wright-Guild Data." *Color Research and Application*, 22(1), 11–23 (1997).—The authors go into detail of how Wright and Guild came to their conclusions which were eventually clarified and adopted by the 1931 CIE in a series of meeting resolutions. This set the standard for calculated conversions between what we know as RGB color values and the theoretical X,Y,Z numerical coordinates which serve as the basis for its color matching system.

34. Fairman, Brill, and Hemmendinger, "How the CIE Color Matching Functions Were Derived."—The authors explain how this problem would have been solved differently today based on the knowledge of human vision that we have now. Their report serves as a critique of several decisions the 1931 CIE made in resolving Wright and Guild's data.

35. Ibid., 13.

36. Margaret Livingstone. *Vision and Art, the Biology of Seeing*. New York, NY: Abrams Books (2002).

37. Ibid., 87.

38. Judd (1966). "Fundamental Studies of Color Vision from 1860–1960."—Judd give a detailed analysis of the issue of non-spectral colors and how the CIE and OSA addressed the issue. He also places this achievement into discussion with the progression and conceptual history of visual theories that shaped the development of color matching technologies.

39. Troland, "Report of Committee on Colorimetry," 553–534.

40. Fairchild, *Color Appearance Models*, 80–82.—Fairchild explains why the CIELAB model has a wider set of applications than the more specific CIELUV, this becoming the standard model.

41. "75 Years of the Colorimetric Standard Observer" (2006).

42. Fairchild, *Color Appearance Models*, 81.

43. Ibid., 79–80.—Fairchild says the CIELAB space is almost universally used as a measurement for color difference specification and color measurement because of its adaptability to a two-dimensional Cartesian plane. The CIELAB color formula functions as a method of constructing a 2D space from 3D data of the L* (light to dark) a* (red to green) b* (blue to yellow) chromaticity coordinates.

44. Buolamwini and Gebru (2018) note the facial recognition feature in a growing list of digital phone platforms like Apple, Google, IBM and Microsoft.

45. A. Kumar and S. Malhotra, "Real-time human skin color detection algorithm using skin color map," *2015 2nd International Conference on Computing for Sustainable Global Development (INDIACom)*, New Delhi, 2015, pp. 2002–2006.

46. Kumar and Malhorta explain their process of capturing skin color data from still and video images in real time using color space and compensating for machine bias variable.
47. Kumar and Malhorta.
48. Benjamin, *Race after Technology*, 36–44.
49. Tiernney, 3.
50. Slack and Wise, *Culture and Technology*, 35–38.
51. Hughes, 112.

BIBLIOGRAPHY

Benjamin, Ruha (2019). *Race after Technology*. Medford, MA. Polity Press, Print.

Bunge, M. (1966). Technology as Applied Science. *Technology and Culture, 7*(3), 329-347. doi:10.2307/3101932

Buolamwini, Joy and Timmit Gebru (2018). "Gender Shades: Intersectional Accuracy Disparities in Commercial Gender Classification." *Proceedings of Machine Learning Research* 81:1–15, Print

Danks, David and Alex J. London (2017). "Algorithmic Bias in Autonomous Systems." in *Proceedings of the Twenty-Sixth International Joint Conference on Artificial Intelligence*, International Joint Conferences on Artificial Intelligence, 4691–4697.

Fairchild, Mark D. (2005). *Color Appearance Models*. Hoboken, NJ: John Wiley and Sons, Print.

Garvie, Clare, Alvaro Bedoya, and Jonathan Frankie (2016). "The Perpetual Line-Up: Unregulated Police Face Recognition in America." *Georgetown Law Center on Privacy and Technology*. Georgetown Law's Center on Privacy & Technology. DOA. 5/1/20 Web. https://www.law.georgetown.edu/privacy-technology-center/publications/the-perpetual-line-up/

Gillespie, Tarleton (2014). "The Relevance of Algorithms." In *Media Technologies: Essays on Communication, Materiality, and Society*. ed. Tarleton Gillespie, Pablo Boczkowski,and Kirsten Foot. Cambridge, MA: MIT Press.

Green, Phil (1999). *Understanding Digital Color*. Sewickley, PA: GAFT Press, Print.

Guild, J. (1932). "The Colorimetric Properties of the Spectrum." *Philosophical Transactions of the Royal Society of London* 230: 149–187.

Heesen, Remco (2015). The Young-(Helmholtz)-Maxwell Theory of Color Vision [Preprint]. http://philsci-archive.pitt.edu/id/eprint/11279 (accessed 2015-09-20).

Helmholtz, H. (1855). Über die Zusammensetzung von Spectralfarben. *Annalen der Physik* 94: 1–28.

"History of the CIE 1913-1988" Photocopy Edition (1999). Vienna, Austria: Comission Internationale de L'eclairage, CIE Central Bureau. Web. Retrieved on 7/22/2017. WWW: http://www.cie.co.at/

Hughs, Thomas P. (1994). "Technological Momentum." In Leo Marx and Merritt Roe Smith (eds.), *Does Technology Drive History?* Cambridge, MA. MIT Press,101–113

Jones, Loyd A. (1943). "The Historical Background and Evolution of the Colorimetry Report." *Journal of the Optical Society of America A* 33(10): 534–543.

Judd, D. B. (1966). "Fundamental Studies of Color Vision from 1860–1960." *Proceedings of the National Academy of Sciences of the United States of America* 55(6): 1313.

A. Kumar and S. Malhotra, "Real-time human skin color detection algorithm using skin color map," *2015 2nd International Conference on Computing for Sustainable Global Development (INDIACom)*, New Delhi, 2015, pp. 2002–2006.

Lee, Barry B. (2008). "The Evolution of Concepts of Color Vision." *Neurociencias* 4(4): 209–224.

Livingstone, Margaret. (2002). *Vision and Art, the Biology of Seeing*. New York, NY: Abrams Books, 2002, Print, pp. 87–91.

Maxwell, J.C. (1855). Experiments on Colour, as perceived by the Eye, with remarks on Colourblindness. *Transactions of the Royal Society of Edinburgh* 21: 275–298.

Maxwell, J. C. (n.d.). On the Theory of Colours in relation to Colour-Blindness. The Scientific Papers of James Clerk Maxwell, Cambridge University Press. pp 119–125. doi:10.1017/cbo9780511698095.009

Maxwell, James Clerk (1890). "Colour vision." In W. D. Niven (ed.), *Scientific Papers Vol. 2*. Cambridge, England: Cambridge University Press, pp. 267–285.

Selena, Silva and Martin Kenney (2019). "Algorithms, platforms, and Ethnic Bias: A Diagnostic Model." Communications of the ACM, November 2019, Vol. 62 No. 11, 37–39.

Selena, Silva and Martin Kenney (2018). "Algorithms, Platforms, and Ethnic Bias: An Integrative Essay." Phylon: The Clark Atlanta University Review of Race and Culture 55(1 & 2): 9–37.

Slack, Jennifer Daryl and Macgregor Wise (2005). *Culture and Technology: A Primer*, 2nd edition. New York, NY: Peter Lang, Print.

Striphas, Ted (2015). "Algorithmic Culture." *European Journal of Cultural Studies* 18(4–5): 395–412.

——— (2019). "Surveillance Giants: How the business Model of Google and Facebook Threatens Human Rights." *Amnesty International*. Web. www.amnesty .org.

Tierney, Thomas F. (1993). *A Genealogy of Technical Culture: The Value of Convenience*. Albany, NY: University of Albany Press, Print.

Troland, L. T. (1922). "Report of Committee on Colorimetry." *Journal of the Optical Society of America and Review of Scientific Instruments* 6(6) 527–596.

Wright, W. D. (1928). "A Trichromatic Colorimeter with Spectral Primaries." *Transactions of the Optical Society* 29(5), 225–242.

Wright, W. D. (1929). "A Re-Determination of the Trichromatic Coefficients of the Spectral Colours." *Transaction of the Optical Society* 30(4): 141–164.

"75 Years of the CIE Standard Colorimetric Observer" (2006).

Chapter 9

Generative Adversarial Networks

Contemporary Art and/as Algorithm

James MacDevitt

Within the past fifty years, all manner of complex procedures traditionally reserved for human agents have been systematically converted into sets of coded instructions to be performed instead by computational machines. Sequentially executable directives—otherwise known as algorithms, after the ninth-century Persian polymath Muḥammad ibn Mūsā al-Khwārizmī—were clearly in existence long before the current era of advanced electronic computers (cooking recipes are the most common and colloquial of examples), but the practical integration of computational machines into virtually all aspects of contemporary human activity has had the simultaneous effect of elevating the automated algorithm into a position of significant, if not yet supreme, shaper of human experience. Algorithms no longer just metaphorically replicate, or even functionally replace, human actions; they now indirectly influence, and in some cases even directly supervise, those activities as well, often without the conscious knowledge of the human beings involved. Ted Striphas, following Alexander Galloway's earlier use of the term, has labeled this general phenomenon "algorithmic culture," noting the myriad ways human beings are increasingly delegating "the work of culture—the sorting, classifying, and hierarchizing of people, places, objects, and ideas—to data-intensive computational processes."[1] Though still a relatively recent development, this overarching trend has only accelerated exponentially in a time of ubiquitous computing connected through high-speed Internet routers, mobile devices, and the so-called Internet of Things; not to mention server clusters with ever-growing processor speeds, access to massive arrays of networked storage capacity, and increasingly comprehensive databases of both natively digital and recently digitized cultural artifacts. In fact, it is precisely this growing ubiquity of algorithmic processing within our contemporary lived experience, much of it now obscured behind curtains of proprietary code and strict architectures of control, resulting in a sustained sense of inscrutability that Warren Sack labels the "computational condition,"[2] which makes unraveling the increasingly interwoven histories of culture and computation so timely and important.

Beyond just sorting and classifying, however, algorithms are also now, more than ever, directly involved in the production of cultural objects themselves (from the innocuous automated response suggestions in text messaging applications to the spectacular code-generated visual effects in big-budget blockbuster movies). While generally understood as a fundamental—perhaps, even uniquely defining—human capability, artistic production has proven itself to be no less susceptible to the creeping influences of algorithmic processing than any other human endeavor, often leading to complicated existential questions regarding the locus of creative imagination, the nature of agential intent, and even the very definition of Art itself. Recent innovations enhancing the efficacy of artificial intelligences—most notably the popular technique for machine learning developed in 2014 by Ian Goodfellow, known as generative adversarial networks (GANs)[3]—have put a renewed spotlight on the common use of algorithmic procedures in the creative production of visual

Figure 9.1 Obvious (Hugo Caselles-Dupré, Pierre Fautrel, and Gauthier Vernier), *Portrait of Edmond de Belamy*, 2018. Ink. 27.5 × 27.5 inches. *Source*: Courtesy of the artists.

imagery, once thought to be the sole domain of human agents, and particularly gifted ones at that. In late 2018, for example, the well-known auction house *Christie's* publicized the sale of an unusual artwork, *Portrait of Edmond de Belamy* (Figure 9.1), for $432,500, well above the initial asking price. What made this sale so notable, according to *Christie's*, was the artwork's status as a completely original figurative image conceived and created by an artificial intelligence (AI) instead of by a human artist. To emphasize this fact, the work was signed with the algorithmic formula used to develop that AI rather than by the actual human team that initiated the project: a French collective known as Obvious, made up of Hugo Caselles-Dupré, Pierre Fautrel, and Gauthier Vernier.

In an interview following the sale, Caselles-Dupré explained the process for producing the work, which is effectively a specific iteration of the now ubiquitous GAN technique used for general machine learning and especially popular for new image generation:

> The algorithm is composed of two parts. On one side is the Generator, on the other the Discriminator. We fed the system with a data set of 15,000 portraits painted between the 14th century to the 20th. The Generator makes a new image based on the set, then the Discriminator tries to spot the difference between a human-made image and one created by the Generator. The aim is to fool the Discriminator into thinking that the new images are real-life portraits. Then we have a result.[4]

That resulting image, to this human's eye, looks simultaneously foreign and yet still somehow remarkably familiar. The familiarity almost certainly stems from the way the AI's machine learning algorithm synthesized, and therefore replicated, the fairly standardized tradition (costuming, body positioning, facial expression, etc.) of painted aristocratic portraiture in European art history (and an obvious cultural, and even class, bias inherent in the data set should definitely be noted here). Whereas the corresponding sense of strangeness is likely a byproduct of the choice to ingest, and therefore amalgamate, images representing over six hundred years of stylistic variations in applied painting technique.

As Adrian Mackenzie has compellingly argued in *Machine Learners: Archeology of a Data Practice*, it is rarely possible "to disentangle machine learners from the databases, infrastructures, platforms, or interfaces they work through."[5] Without a trained art-historical appreciation for the evolutionary character of stylistic changes within the last half-millennium of painting techniques, the algorithm simply averages these very particular and nuanced aesthetic qualities together. While the resulting image may appear interesting, even appealing in its aesthetic flourishes, its character is also definitively ahistorical. The work as a whole, then, cannot, and should not, be merely appended to the existing genealogical narratives that frame modern and contemporary painting, which may, at least, be of some comfort for those who feel the knee-jerk fear often engendered by this kind of AI-produced imagery (i.e., a fear that the machine will eventually eclipse, and ultimately replace, the need for a human "artist/author"). Of course, it could easily be argued that, in the long run, this was effectively an early proof-of-concept prototype and that a certain art historical awareness of stylistic chronology could be programmatically incorporated in the future with additional machine learning code and more refined data sets, which is undoubtedly feasible on a technical level (see, for example, *The Next Rembrandt* project).[6] But this might also be missing the point entirely. Creating evermore convincing facsimiles of human-made artworks by artificially intelligent agents, whether they seek to replicate the stylistic flourishes of specific human artists (i.e., the increasingly popular neural style transfer algorithms) or to embrace the uncanny surrealism of GANism itself, will quickly lose its charm and cease to be *conceptually* interesting, even if they continue to produce compellingly original *formal* innovations (just look at how quickly once groundbreaking algorithms have devolved into overly used and clichéd video filters in social media applications).

In fact, the most notable aspect of *Portrait of Edmond de Belamy* was not its formal, or even its technical, originality (both of which have, rightly or wrongly, received substantial criticism already).[7] It was, rather, the choice to sign the work with a portion of the algorithmic formula used to produce it, visually marking the author/artist as the AI itself, and then circulating said work within an institution that traditionally functions as one of a number of powerful gatekeepers to the collective cultural and economic valuation of fine art objects. Both of these important decisions were undertaken by the collective known as Obvious, and not the AI platform they built. The formal decisions might be a result of the AI and its underlying code, but the conceptual innovation and contextual locus of its reception were still very human. Pointing this out is not meant to take anything away from the significance of the work, nor is it meant to participate in that rather tired and overly romanticized argument about the uniqueness of the human *qua* artist and vice versa, where "the concepts of art and inspiration are often spoken of in mystical terms, something special and primal beyond the realm of science and technology; as if only humans create art because only humans have 'souls.'"[8] Rather, at some point, whether or not a work was *actually* created solely by an AI is less important than whether or not we *identify* it as such.

Situating this work outside the current, perhaps overly profusive, narratives surrounding ever perfecting algorithms, machine learning, and AI, and instead acknowledging its debt to existing, and quite analog, art historical movements, such as Conceptual Art and Institutional Critique, simply allows for the possibility of further theoretical connections to be made and deeper analyses to be pursued than could otherwise be accomplished through a purely formal analysis alone. Going down this road, for example, one might instead link Obvious's seemingly blasphemous insertion of an artificially created object into a fine art institution with the historic precedent of Marcel Duchamp's most notorious readymade, *Fountain*, a simple urinal that he purchased and signed as a work of art almost exactly one hundred years prior to the creation of *Portrait of Edmond de Belamy*; whereas Duchamp's piece could be seen as a statement about the changing status of Art during the Machine Age, Obvious's work might be understood as a similar response to the artistic challenges of the Information Age. Or, invoking the theoretical writing of Jacques Derrida—in particular his deconstructive text, *Glas*[9]—one might critically explore signatures as physical marks of authorial identity, existing both within and outside of the "frame" of the work of art itself; in the case of *Portrait of Edmond de Belamy*, the signature identifies the AI as author (perhaps misleadingly), but also, by revealing the exact algorithmic formula used, marks the work itself as both process and product. Even the title of the piece, which references the fictitious Belamy family, could be read as a consciously playful invocation of the AI as artist; Belamy is phonetically equivalent to the French term *bel ami*, which roughly translates to "good friend," a possible allusion to Ian Goodfellow, the original inventor of the GAN process, or, alternately, could be seen as a reference to the late nineteenth-century French novel *Bel Ami* by Guy de Maupassant, the story of a "scoundrel" artificially rising to prominence through luck and a fair bit of social manipulation and maneuvering.[10]

Whether any of these associations were actually intentional or not, be it on the part of the AI or its human creators, is probably irrelevant. The broader point is this: what often gets lost in the too-frequent and overly simplistic debates about whether or not a machine, a computer program, and now an AI, can "make art," is that much of our human appreciation of any specific artwork actually comes from the stories we tell about it which is decidedly *not* a part of the so-called creative process. It is, instead, a defining aspect of reception. From that perspective, external contexts matter as much as, if not more than, the art object itself up to, and including *après la lettre* the discursive assignment of authorship, as Michel Foucault posited in "What is an Author?," his 1969 lecture at the Collège de France:

> Nevertheless, these aspects of an individual, which we designate as an author (or which comprise an individual as an author), are projections, in terms always more or less psychological, of our way of handling texts: in the comparisons we make, the traits we extract as pertinent, the continuities we assign, or the

> exclusions we practice. In addition, all these operations vary according to the period and the form of discourse concerned.[11]

Think about all the recent problematic calls to separate previously celebrated artworks from their since tainted artists, as if such a thing was even possible, in the wake of the #MeToo movement or, long before that, and in direct opposition, the extensive research conducted by numerous feminist art historians to recover the long lost/erased work of forgotten/ignored women artists. If an artwork existed solely in a vacuum—if it *was* only a synthesis of formal elements changing over time, which is effectively the conception of "Art" programmatically coded into many of the popular image generating machine learning algorithms circulating today—then the identity of the artist would actually not matter to the appreciation of the work itself, in which case a human or a machinic or an AI creator *would* be all but interchangeable.

However, as Andreas Broeckmann has argued in "The Machine as Artist as Myth," it is, actually, the very definition of 'Art' itself that has been changing over time, and in no small part exactly because of the technological changes that have impacted society coeval with shifts in the practice of artmaking:

> The fact that art is done by an artist is as essential an aspect of modernist and postmodernist art as the very question, "What is Art?" From Abstract Expressionism through Pop Art to Institutional Critique, discussions about art have been a pivot for debates on human production and creativity in the age of mass industrial production and consumption. Questioning the status of the artist is an inherent part of these considerations ever since the non-sense performances of Dada, the psychic automatisms of Surrealism, and the mathematical automatisms of Concrete Art.[12]

And, just as Foucault pointed out that the concept of authorship is itself not universal, but historically constituted, such that a culturally mandated authorial requirement differs for different kinds of texts at different times, he also acknowledged that the shifting necessity for an assigned author (a human author is here implied) has technically been a matter of adjustments to collective legal codes more than any grand humanist celebration of imagination or individuality. "Speeches and books were assigned real authors, other than mythical or important religious figures, only when the author became subject to punishment and to the extent that [their] discourse was considered transgressive."[13] Likewise, the status of the visual artist in Western art history has largely been a product of legal codes, from the guild contracts of Gothic-era cathedrals to the 1878 libel case between John Ruskin and James Abbott McNeill Whistler, two examples which largely bookend the epistemic transition from nearly anonymous craftsperson to completely individuated experimental modern artist, generally assumed to use their "vision" to depict the world as no one had done before and their "voice" to comment upon existential dilemmas and societal woes alike.

Long in the making, this now familiar, though no less historically construed, variation of the "solo artist as visionary creator and social critic" (again, human was always implied, as were, until recently, white and male, for that matter) was the one that reigned supreme around the time that some of the earliest examples of algorithmically based computer-generated visual imagery was being introduced to the world. In 1965, in response to the first of its kind *Computer-Generated Pictures* exhibition at the Howard Wise Gallery in New York, Bell Labs demanded that their contractual research scientists participating in the exhibition "gain copyright as a way to disassociate the work from the scientific research undertaken at Bell Labs,"[14] since Art was understood, at the time, as too frivolous an endeavor for a serious research center to be pursuing.[15] But, this led to another problem. The Copyright Office at the Library of Congress initially refused to give one of those computer artists, Dr. A. Michael Noll, a copyright for his submitted work, including *Gaussian-Quadradic* (Figure 9.2)—which, it should be noted, he himself chose to identify as "patterns," rather than "Art"—on the grounds that a machine had made the work (of course, they certainly weren't going to award the machine a copyright either, creating an effective standoff). In order to finally register the work under his own name, Noll had to repeatedly explain that he had written the program executed by the machine and that, although the numbers generated by the program "appeared 'random' to humans, the algorithm generating them was perfectly mathematical and not random at all."[16] Of course, Duchamp's readymades had probably been the first notable examples within modern art history of work that championed the theoretical notion that the artist need not directly produce a work of art to be considered the "creator" of that piece, but Noll's copyright assignment for *Gaussian-Quadradic* may actually have been one of the first legally binding codifications of that very concept.[17]

In fact, despite being a somewhat marginalized and overlooked aspect of twentieth-century art history, both the early experiments to create computer-based art in the 1960s and the subsequent technological developments since that time have actually been quite significant in shaping new perspectives in what has become contemporary art. As Broeckmann also points out:

> "Art" has been many different things in different cultural and historical contexts. The role of "the artist" in the Modernist art of the 20th century is far more complex than that of an intentionally acting "creative individual," and some of the key issues in this debate—which the applications of artificial intelligence systems now also address—concern aesthetic transformations of concepts,

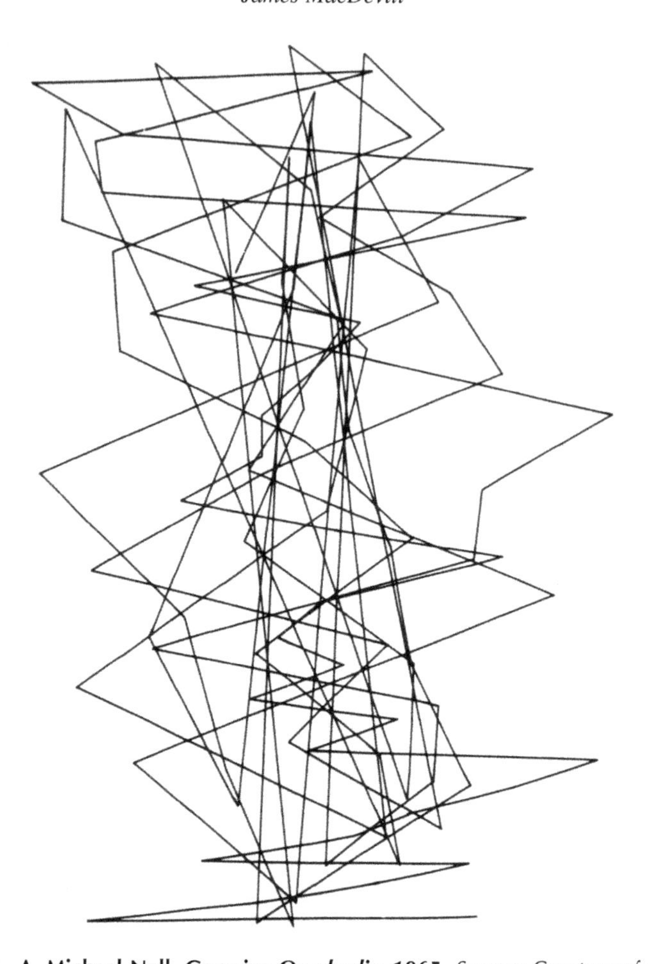

Figure 9.2 A. Michael Noll, *Gaussian-Quadradic*, 1965. *Source*: Courtesy of the artist.

processes, and systems that have been addressed extensively in the contexts of conceptual art and systems aesthetics, ever since the 1960s.[18]

It's not that exploring whether a machine, a computer program, or an AI can "make Art" isn't an interesting and relevant question. Obviously, from the examples above, many aspects of modern and contemporary art have directly or indirectly attempted to tackle this very issue. But if the line of inquiry stops there, then many other pertinent questions remain unanswered. Rather, as Carolyn L. Kane has suggested in *Chromatic Algorithms: Synthetic Color, Computer Art, and Aesthetics after Code*, what is really needed right now is an analysis that pays "critical attention to the material and ontological processes involved in algorithmic processing"[19] as this will "open up broader questions concerning social and cultural operations: the production of visual knowledge, concerns about privacy, shifts in the political and economic infrastructure, and perhaps most importantly, what it means to be alive, and desiring in the algorithmic lifeworld."[20]

Recognizing the historical embeddedness of specific artworks, as well as the discourses surrounding them, necessarily provides a better gauge to the causal connections between the broader evolution of technologies within society and the critical artistic responses to those changes. For example, Paul Delaroche, a highly respected member of the French Académie des Beaux-Arts, after viewing a demonstration of the new daguerreotype photographic process in 1839, supposedly made the, since then, oft-repeated prediction that "from today, painting is dead." In the face of a largely automated technology that could represent the world at a speed and accuracy impossible for human painters to challenge, Delaroche rightly recognized that the painter's traditional societal function was threatened, if not doomed. What he clearly was unable to imagine at the time, however, was that painting would not actually die in response to the invention of photography. Instead, its mission statement would have to be rewritten in order to maintain cultural relevancy, with a significant shift in focus away from traditional modes of naturalistic representation toward the more formal and stylistic innovations that we today associate with modernism. From photography to the assembly line to computers to AI, "in each case, we see that these new technologies caused fears of displacing artists, when, in fact, the new technology both created new opportunities for artists while also invigorating traditional media."[21] Additionally, in each case, the most forward-thinking and, ultimately influential, artists have generally been the ones that saw these changes coming and adapted their art practices accordingly.

Not surprisingly then, far from the sentimental view of art that was clearly still held by the public relations team at Bell Labs in 1965, the most interesting and innovative artists of that same era were already deep into exploring the possibilities and implications of algorithmic and code generated systems. A procedural epistemology was clearly at work for Sol LeWitt when he claimed in his manifesto, "Paragraphs on Conceptual Art," first published in the June 1967 edition of *Artforum*, that "in conceptual art the idea or concept is the most important aspect of the work. When an artist uses a conceptual form of art, it means that all of the planning and decisions are made beforehand and the execution is a perfunctory affair. The idea becomes a machine that makes the art."[22] It was this kind of thinking that would later lead LeWitt to produce his famous *Wall Drawings*, in which different manifestations could be physically inscribed onto the gallery walls by surrogate artists or museum preparators, while the work itself was always considered the series of specific instructions informing the creation of those manifestations designed and penned by LeWitt himself. Effectively, LeWitt wrote the program that generated the work or, in his words, "the idea becomes the machine that makes the art."[23]

And he was not alone in building artistic scripts to be executed by others during the 1960s. For example, at the beginning of that decade, in July of 1961, the young Fluxus artist, Yoko Ono, exhibited a series of paintings at George Maciunas's AG Gallery in New York which included hand-written instructions for how to interact with the pieces.[24] One of those artworks, *Painting to Be Stepped On*, consisted of a piece of cut canvas on the floor with a notecard next to it that read "A WORK TO BE STEPPED ON." A year later, at the Sogetsu Art Center in Tokyo, Ono exhibited a related series of new paintings that consisted solely of instructions written in Japanese calligraphy describing how to recreate many of those earlier works. So, for the second iteration of *Painting to Be Stepped On*, the words simply stated, "leave a piece of canvas or finished painting on the floor or in the street."[25] It was up to the viewer to follow through on the instructions, if they so desired. No objects were displayed this time, other than the calligraphic instructions, which themselves were actually drawn not by Ono, but by her then-husband, Toshi Ichiyanagi, a former student of the experimental musician, John Cage.[26]

Cage had actually been one of the very few visitors to the opening reception of Ono's show at AG Gallery, where she debuted her conceptual work for the first time.[27] Ono, like many other Fluxus and Conceptual artists, admired Cage's musical experimentation, including his revelatory use of atypical visual notation to add chance and randomness to his performances, such as his well-known 1958 score for *Fontana Mix*. As Liz Kotz highlights in her book *Words to Be Looked At: Language in 1960s Art*, Cage's approach to musical composition was at least partially formed by his encounter with electronic sound equipment and computational recording technologies. "With the introduction of electric, electronic, and computer-based means into experimental music, by the 1960s, anything from circuit diagrams to punch cards to simple drawings and verbal instructions could arguably function as scores or notational devices."[28] Cage, himself, in his 1940 manifesto, "The Future of Music: Credo," claimed that "the new materials, oscillators, turntables, generators, means of amplifying small sounds, film phonographs, etc., available for use [allow modern composers] to capture and control those sounds, to use them not as sound effects but as musical instruments."[29]

Another one of Cage's students, and an active member of the Fluxus group along with Ono, was George Brecht. As early as 1959, Brecht was creating what he referred to as "card events" which "consisted of small printed instructions that outlined detailed procedures for a variety of loosely synchronized actions."[30] These clearly resembled the card-based instructions that were typically converted to magnetized tape and fed into mainframe computers at the time, both those encoded directly as holes on "punch cards," and, even more literally, the written "instructions on 'coding sheets' (a phrase that gives rise to our current reference to computer programming as coding) that were then given to keypunch operators to create the cards."[31] For Brecht, the codes he used to orchestrate his events were both physical and situational. They could include "raising and lowering the volume of radios, changing the tuning, and so forth, for indeterminate durations based on natural processes such as the burning of a candle (*Candle Piece for Radios*, summer 1959); or turning on and off various lights and signals, sounding horns, sirens, or bells, opening or closing doors, windows, or engine hoods, and so on (*Motor Vehicle Sundown [Event]*, spring-summer 1960)."[32]

Underlying much of Brecht's—as well as LeWitt's, Ono's, and Cage's—approach to art-making was a desire for indeterminacy; a fascination with "random order" and "permutational systems" as a means of avoiding both repetitive habits and overly sentimental expression.[33] "This drive to escape habit and cliché led artists [like these] into considering algorithms and system as a means to conjure a different mode of art that sidestepped questions of intention."[34] So, the mid-century embrace of algorithms by artists was largely due to an interest in giving up control, seeing in aleatory and stochastic systems a randomness that mirrored the unpredictability of the world itself. For their contemporaries in the newly minted discipline of the computer sciences, however, the embrace of algorithms was initially far more practical. Algorithms were historically understood among mathematicians as "step-by-step procedures designed to solve well-defined problems, [specifically] the operations of pen applied to paper to perform calculations using the then-new system of Hindu, decimal, fixed-point arithmetic."[35] As Warren Sack has shown in his recent book, *The Software Arts*, it was only much later that algorithms "came to have a more expansive meaning—beyond simple arithmetic—after the arithmetization of mathematics and the development of theoretical and then electronic computers"[36] In fact, as Sack also points out, the term "algorithm" was not explicitly associated with the composition of computer software until around the 1960s, when Donald Knuth published his influential textbook *The Art of Computer Programming*, a multivolume work (still being expanded today), which focuses its first couple volumes on a detailed analysis of both theoretical and specific algorithms.[37]

However, the use of algorithms in computer programming was never just a simple matter of mathematical calculations. It was also, and even primarily, an issue of logistics; using various specialized languages to describe how some particular isolated action should be completed within a broader sequence of events. As such, the origin of computer programming, as we know it today, could be understood as not just as an extension of the discipline of mathematics, but also emergent from developments in the philosophy of symbolic logic and, in an indirect fashion, from the Enlightenment era literary publication of the comprehensive *Encyclopédie* by Denis Diderot and Jean le Rond d'Alembert, which included detailed descriptions of the multiple procedures and divisions of labor used in the manual and mechanical arts to produce various cultural and technical objects.[38] As design historian Antoine Picon explains:

> The common threads that connect the different articles devoted to the arts and crafts are the description of elementary gestures of production, how these movements are integrated and thereby define aggregate technical operations, and the logic of chaining together these operations to form processes organized according to a division of labor. . . . From individual movement to process chain, the thread that weaves them together is analogous to the overall aim of Diderot, d'Alembert, and their *Encyclopédie* collaborators: the integration of all forms of knowledge.[39]

In other words, summarizing Warren Sack's more lengthy and brilliant argument linking the encyclopedists to the origins of algorithmic thinking in his book *The Software Arts*, "the work language of the arts anticipates what we know today as computer programming languages."[40]

This is largely because, before computers were assumed to be machines, they were actually people. "Computer" was basically a job description ("one who computes"), a laborer not unlike the generic "Pinmaker" described in the *Encyclopédie* (later co-opted by Adam Smith in *The Wealth of Nations* as his quintessential example of the division of labor, one of the keystones in his argument for the efficiencies of economic capitalism). Human computer was far from a glamorous profession, however; it was tedious and considered undesirable, which may have been why it was often assigned to women, such as Elizabeth Langdon Williams and Henrietta Swam Leavitt, whose contributions to important scientific discoveries are only now being recognized. Charles Babbage, in designing the prototypical computational machine, his theoretical *Analytical Engine*, in 1837, actually hoped to remove all need for human computers.[41] Ava Lovelace, a talented mathematician, as well as the daughter of the poet Lord Byron, would later publish the first functional algorithm meant to be processed by Babbage's machine, making her, in the eyes of many, the first computer programmer. Part of Babbage's design for the *Analytical Engine* included punch cards to feed input variables and algorithmic logic controls into the machine, a practice that continued well into the next century with the electronic descendants of Babbage's machine, including the same computer mainframes used by the early computer artists. The inspiration for the use of these punch-card style instructions, of course, came from the famous Jacquard loom, designed by Joseph Marie Jacquard in 1804, a practical device for automating the templating patterns of mass-produced woven textiles.

So, both the theoretical precedents for computer software and the literal precedent for computer hardware can be traced directly to the manual production of *objets d'art* and a desire to automate production thereof. Well before the 1960s, culture and computation were clearly connected to each other through numerous overlapping networks. Perhaps the primary nodal connection within those overlapping networks, however, was the shared elevation of the lowly algorithm, laying the foundation for more significant interactions between art and technology in the latter half of the twentieth century and beyond. Of course, highlighting the parallel histories of computer art and conceptual art in the 1960s is not necessarily anything new. Edward Shaken, writing in *Leonardo* in 2002, claimed that "the correspondences shared by these two artistic tendencies offer grounds for rethinking the relationship between them as constituents of larger social transformations from the machine age of industrial society to the so-called information age of post-industrial society."[42] And Christine Tamblyn, in her 1990 essay for *Art Journal*, "Computer Art as Conceptual Art," pointed out that the then-new computer-based and net.art experiments of artists such as Lynn Hershman (Leeson), Nancy Burson, and Christopher Burnett owed much of their particular approach to earlier Conceptual artists. Not that all of this wasn't obvious to anyone paying attention. A simple side-by-side comparison of Manfred Mohr's 1973 computer-generated plotter-drawing *P-154C* and Sol LeWitt's 1974 conceptual installation *Incomplete Open Cubes* is really all it takes to recognize the similarities between these two artists and the broader movements they represented (clean linear geometry, fragmentation and reassembly, serialization of forms, etc.). Although, as Grant D. Taylor has pointed out in his phenomenal study, *When the Machine Made Art: The Troubled History of Computer Art*, there were still subtle differences based on the particular interests and agendas of the two artists. "In LeWitt's work, the tension of the work arises from the relationship between the idea and its physical realization. For Mohr, in contrast, it resides in the potential of the computer algorithm and its power to generate vast amounts of signs."[43]

Less explored, however, is how developments in computer-generated art and algorithmic conceptualism continued to actively interact in the decades that followed. How do we get, for example, from Mohr's *P-154C* and LeWitt's *Incomplete Open Cubes* to a visually analogous project like *In Search of an Impossible Object* (Figure 9.3), produced nearly half a century later by Philipp Schmitt and the Designed Realities Studio of Anthony Dunne and Fiona Raby? Though the aesthetic corollaries are heavy, *In Search of an Impossible Object* ultimately has more in common technically with *Portrait of Edmond de Belamy* than with anything made by Mohr or LeWitt. Similar to Obvious's GAN-generated painting, *In Search of an Impossible Object* was produced by training a machine algorithm using a generative

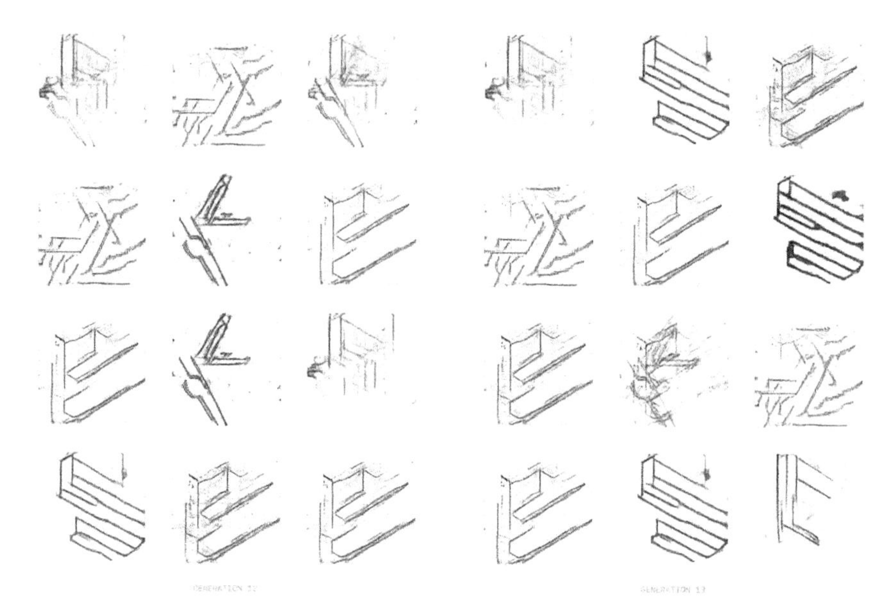

Figure 9.3 A project by Dunne & Raby and Philipp Schmitt for the Designed Realities Studio, *In Search of an Impossible Object (Generation 13)*, 2018. *Source*: Courtesy of the artists.

adversarial network. Instead of European portraiture, however, the data set here consisted of known images that produce optical and spatial illusions for the human eye. Once trained in the basic characteristics of the rather limited number of examples of optical and spatial illusions currently available, the team at the Designed Realities Studio coaxed the AI to produce a new illusionary example, but to no avail. Instead of featuring a single successful image, the Designed Realities Studio created an entire book revealing each of the many failed attempts by the AI to generate a unique illusionary form. Per the artists' description:

> Each page shows a selection of attempts to visualize an impossible object. None succeed. But collectively, they suggest something lying just beyond what can be represented graphically. It's as if a machine, for a moment, attempted to show to humans in a visual language we understand, what is permanently unknowable to us.[44]

The serialized layout of multigenerational permutations pays obvious tribute to both the conceptual and computer artists that came before; this is a work that clearly is aware of the history in which it swims. But the ideation behind the work looks directly at the complexities of living in the present moment and the tensions that permeate our "algorithmic lifeworld." It's as if, for just a second, those curtains of code pull back, revealing the failings of the artificially intelligent wizard hiding within.

The criticality of this approach is itself a product of how art and technology have intersected and interacted within the intervening decades between the 1960s and the present. These shared and overlapping histories have served as their own kind of generative adversarial network, with each perspective challenging and perturbing the other; technological innovators forever trying to generate aesthetically original imagery that might be accepted within traditional art contexts and contemporary artists either appropriating the structural order and visual lexicons of algorithmic computation or working to expose the conceptual pitfalls of these technological systems, which increasingly happens these days in real time as the artists frequently work through, and harness the possibilities of, the very systems they are critiquing. Evolutionary changes in algorithmic computation and developments in contemporary art are by now completely intertwined, each borrowing liberally from, and/or openly decrying, the other, a process that has only improved the aesthetic and ethical capabilities of each. Tracing both of these parallel, occasionally confrontational, trajectories—and laying out each step, each permutation, as did Mohr, LeWitt, and Schmitt in the examples above—can help to expose both the generative possibilities and the adversarial agendas of contemporary art and algorithmic culture. Such, in fact, will be the algorithmic process guiding the remainder of this discussion.

Algorithmic computation in the early to mid-1960s was highly influenced by the then dominant authority of modernism as an exemplar of sophisticated artistic style. Early pioneers, initially coming out of the computer sciences rather than the fine arts, searched for validating correlations between the images they were generating using code and the well-known work of established modernist artists. In 1965, the German computer artist Frieder Nake produced a plotter-drawing titled *13/9/65 Nr. 2*, also known as his *Homage à Paul Klee*, due to the visual similarities between this work and Klee's 1929 painting *High Roads and Byroads*. Back in the United States, A. Michael Noll had similarly argued that his work *Gaussian-Quadratic* resembled the linear angularity of Pablo Picasso's famous analytical cubist painting from 1911 to 1912, *Ma Jolie (My Pretty Girl)*. This visual analogy inspired Noll to pursue his so-called Mondrian Experiment, in which he coded a computer-generated semi-replica of Piet Mondrian's *Composition in Line* from 1916 to 1917. Though the placement of the individual marks was generated according to a random number routine, their shape was a product of "statistics chosen to approximate the bar density, lengths, and widths in the Mondrian painting."[45] The resulting work, *Computer Composition with Lines*, was then shown to 100 Bell Labs employees alongside a xerographic copy of Mondrian's original painting. Test subjects were asked to rate which work they liked best and which one they thought was made by a machine. Surprisingly to some, a majority of the test subjects could not correctly identify the original painting, though a clear plurality actually preferred the computer-generated work over the original. This may be because many modernist visual movements—including the Cubists, the Bauhaus, and De Stijl—were themselves shaped by a particular interest in a machine-based aesthetic.

That initial impulse among early computer art researchers to replicate the aesthetics of existing artistic trends was rather quickly replaced, however, by the introduction of trained visual artists into the mix. These artists tended to be more interested in absorbing the visual vocabulary that was specific to algorithmic computation and then using it to develop whole new sets of aesthetic possibilities and experiences. In the United States, the most prominent example of this was the collaborative Experiments in Art and Technology program, which grew out of the professional relationship between the artist Robert Rauschenberg and Bell Labs engineer Billy Klüver. A series of notable E.A.T. affiliated and E.A.T. adjacent exhibitions soon followed, including *9 Evenings: Theater and Engineering* at the 69th Regiment Armory in New York, *The Machine as Seen at the End of the Mechanical Age* at the Museum of Modern Art in New York, *Cybernetic Serendipity* at ICA in London, *Software* at the Jewish Museum in Brooklyn, and the Pepsi Pavilion at *Expo '70* in Osaka, Japan. While not directly associated with E.A.T., the positive cross-disciplinary attitude represented by this ongoing collaboration was also mirrored in the budding willingness at Bell Labs to establish a number of extended artist residencies. Using Kenneth Knowlton's BEFLIX programming language, itself an extension of FORTRAN, artists like Stan VanDerBeek (*Poemfield No.1*, 1966) and Lillian Schwartz (*Pixillation*, 1970) made innovative computer-generated abstract and textual animations that were a cross between the visual languages of experimental film, psychedelic culture, and a pixelated aesthetic that would soon become associated with early video game consoles.

Early computer art in Europe tended to be far more sedate, at least initially, by comparison. Influential pioneers such as Georg Nees and Frieder Nake were students of the philosopher Max Bense, a co-founder of the Stuttgart School at the Technische Universitat, who championed a mathematically based approach to what he called "generative aesthetics." His 1965 treatise, *Aesthetica*, was one of the first texts to theorize the possibility of using computers to "program the beautiful."[46] Inspired by this concept, many computer artists in Europe designed and coded aleatory algorithms which allowed them to explore "entire families of forms."[47] Nees's 1965 plotter-drawing *23-Corner Graphic*, for example, used "random parameters and generative functions" to create a seemingly endless array of unique geometric forms with certain shared visual traits. Building further on Bense's generative aesthetics and its connection to semiotics, specifically exploring the "relationships between signs and systems,"[48] Mohr designed series of "graphic entities" that resemble a "constellation of hieroglyphic-like marks varying in small increments laid out on a matrix."[49] The matrixial grid-like structure that underlays much early computer art was both a perfect framework for the serialized permutations popular among this group of artists, as well as a further validating link back to many of the rationally focused modernist art movements, such as Constructivism and De Stijl, that had also gravitated to this particular form. The European artist who played most liberally with the modernist linear elements of the square and grid was the Hungarian-French artist, Vera Molnár. Coming out of the New Tendencies movement and a co-founder of the radically experimental collective known as GRAV (*Groupe de Recherche d'Art Visuel*), Molnár was herself less enamored with Bense' rational mathematics and preferred to explore forms intuitively through what she called "small probing steps."[50] Through the slight alteration of particular parameters, the orderly modernist grid would slowly collapse upon itself, as is seen in works such as her *(Dés)Ordres* from 1974.

From the geometric orderliness of A. Michael Noll to the formalized deconstructions of Vera Molnár, all of these early pioneers of algorithmic computation would set the stage for, and continue to participate in, the code-based generative art of the next few decades. In the late 1980s and 1990s, with new programs like Adobe Photoshop available for home computers allowing visual artists to interact with digital media without the specific use of programmatic code, the term "computer art" slowly lost its meaning. Artists such as Roman Verostko and Jean Pierre Herbert, both of whom had been working with algorithms for some time, as well as a new generation of artists working with algorithms and code, embraced the alternative term "Algorists" to describe their overarching practice. Herbert even wrote a manifesto using a code-based style reminiscent of the programming languages the Algorists were using to compose their work:

> if (creation && object of art && algorithm && one's own algorithm) {include * an algorist *} elseif (!creation || !object of art || !algorithm || !one's own algorithm) {exclude * not an algorist *}[51]

The coded manifesto makes one thing immediately clear. To be considered an Algorist, one must "uses one's own algorithms for creating art objects."[52]

Direct coding, as opposed to the use of pre-packaged commercial applications, remains a significant element of artists working with generative art today. The limitations of earlier programming languages have even inspired some code-based artists to build new languages from the ground up. In 2001, for example, artist Casey Reas, along with Ben Fry, created the Java-based Processing open-source graphical library and, more recently, Zach Lieberman helped design the OpenFrameworks library, a similar open-source C++ toolkit that, according to their website, "makes it much easier to make things with code."[53] Processing, in particular, with a nearly twenty-year history, has become ubiquitous with creative coders. Its power stems from its flexibility, allowing for artists to easily produce works for print, for screen, for the web (via a javascript-based extension), and for integration with other physical computing elements (via the related Arduino library). Reas's own *Process* series is a prime example of this flexibility, with individual *Process* implementations, what Reas calls "a kinetic drawing machine," presented across multiple media environments from print to screen to installation. The series also serves as an excellent example of the importance the Algorists, and those that have since followed, place on surfacing the underlying algorithms in their work.

In addition, however, Reas also demonstrates a clear inheritance from the conceptual approach of LeWitt, even alluding to LeWitt's famous description of the idea as "the machine that makes the work" by describing his abstract ideated "elements" as machines. In illustrating his methodology for building his individual *Process* iterations, Reas states "An Element is a simple machine that is comprised of a Form and one or more Behaviors. A Process defines an environment for Elements and determines how the relationships between the Elements are visualized."[54] For example, *Element 2* takes the Form of a Circle that exhibits both *Behavior 1* ("moves in a straight line")[55] and *Behavior 5* ("entering from the opposite edge after moving off the surface").[56] *Process 8*, which then incorporates *Element 2*, states:

> A rectangular surface densely filled with instances of *Element 2*, each with a different size, speed, and direction. Display the intersections by drawing a circle at each point of contact. Set the size of each circle relative to the distance between the centers of the overlapping Elements. Draw the smallest possible circle as black and largest as white, with varying grays representing sizes in between.[57]

Like a scientist identifying the atomic structure of specific molecular elements, Reas combines his basic Forms with particular Behaviors to assemble his very own periodic table. These molecular Elements can then be let loose in a controlled Process(ing) environment, with the artist the first observer of many. The organic and biological metaphors here are not accidental, as both Reas and many of the generative artists working in the twenty-first century have taken a particular interest in what have become known as "emergent" forms, complex patterns arising from a large number of smaller interactions. A murmuration of birds or a swarm of ants or a network of brain neurons are all real-world examples of familiar emergent forms. Long gone are the simple geometric shapes and linear elements of the early computer artists. In their place are now complex, nuanced arrangements that are generated on the fly, often in real time as animations for a screen. The careful assemblage of repeated layered elements and specific rules guiding the emergent forms in Reas's work allows for his pieces to have, despite appearing in perpetual variations, a rather singularly defining aesthetic and associated authorial identity.

As has happened, however, with other platforms and applications intended to lower the restrictions to access and use, the easy learning curve of coding libraries such as Processing, combined with the growing number of standardized algorithms in regular use, have made for an environment where many generative art pieces are effectively indistinguishable from one another. One well-known generative artist, Marius Watz, highlighted this dilemma, and sparked a raucous debate, when he posted a tweet challenging his generative artist peers to avoid these kinds of algorithmic clichés: "@blprnt We talked about this. Voronoi is off limits until 2015, it got used waaay too much by architects in 2011. Temporarily banned algorithms: Circle packing, subdivisions, L-systems, Voronoi, the list goes on. Unless you make it ROCK, stay away."[58] Watz expanded his thinking in a longer post to his Tumblr page with the headline *The Algorithmic Thought Police*:

> Upon "discovering" an elegant algorithm that yields compelling visual results (say, circle packing or reaction-diffusion) there is a strong temptation to exploit it as is, crank out a hundred good-looking images and post them all over your Flickr, your blogs, what have you. I've done this. If you're reading this you've probably done it too, and you know what happens next. Suddenly you find that the dude/dudette next door "discovered" the exact same algorithm and made a hundred images just like yours. And there's egg all over your face.[59]

Of course, one possible alternative to simply relying upon overly repetitive algorithmic patterns which might incrementally reveal their underlying structure to viewers with each new iterative manifestation is to seek solace in the creative potentiality of AI.

This may be why AI is currently having what is often called "a moment." As generative art has matured and, in the process, become somewhat tiresome in its familiarity, AI has stepped in as the next possible frontier for practical originality in algorithmic computation. Auction houses like *Christie's* have now sold work ostensibly created by AI artists (i.e., *Portrait of Edmond de Belamy*) and an endless number of exhibitions have recently showcased a variety of both overly optimistic and hypercritical examples of AI art. Just within the last couple years, there has been Refik Anadol's *Machine Hallucination* at Artechouse, *Unsecured Futures: The Art of Ai-Da Robot* at the Barn Gallery at St. John's College, *Faceless Portraits Transcending*

Time: AICAN + Ahmed Elgammal at HG Contemporary, *Uncanny Valley: Being Human in the Age of AI* at the DeYoung Museum, and *The Question of Intelligence: AI and the Future of Humanity* at The New School's Sheila C. Johnson Design Center. Of course, AI did not appear out of nowhere. AI was first seriously explored by scientific researchers all the way back in 1956 at the *Dartmouth Summer Research Project on Artificial Intelligence* at Dartmouth College in Hanover, New Hampshire. The proposal for the conference stated that "the study is to proceed on the basis of the conjecture that every aspect of learning or any other feature of intelligence can in principle be so precisely described that a machine can be made to simulate it."[60]

A decade or so later, Harold Cohen, a professor at UC San Diego, was the first to speculate about what exactly it would mean for a machine to adequately simulate a fundamental feature of intelligence such as making art. He argued that a true AI artist would not just recreate an optical representation of the world, à la a photographic camera; instead, the AI would need to actually have some general "understanding" of the world outside itself in order to then depict that world as would a human artist. Any successful artwork, by Cohen's definition, would include "the trace not only of what is seen, but more important, of what is known. They're not merely photographic images of surfaces that strike the eye; they're images that embody some things the image-maker knows about those objects"[61] Accordingly, Cohen's development of an AI that could actually accomplish this goal, a program known as AARON that he continued to refine over the course of nearly forty years, was really a deep dive into the very nature of human cognition and early childhood psychological development. With each new primitive cognition routine Cohen added to the program, the resulting drawings it produced became more complex. He started with outlines, exploring how human brains identify the edges of objects in the world and writing code that would simulate that activity. Later, he added algorithms for the handling of expression and color. Rather than just replicating the output of a brain, he was actually building one from the ground up. His exhibitions were spectacular affairs, as he would set the program running and allow it, through a remote-controlled "turtle," to paint directly onto large pieces of paper or canvases laid out on the floor. Visitors would get to experience not just the end point of the AI-created work, but the "creative" process that led to that result.

More recently, the artist and roboticist, Pindar Van Arman, has used machine learning AIs to control robotic arms that swipe brushes across the surface of canvases to create original paintings. Recognizing that the layered materiality of the paint contributes to the expressive nature of the medium and attempting to avoid the repetitive clichés of other generative art practices, Van Arman has created an AI known as *CloudPainter*, which combines multiple successive algorithms to make as many independent decisions as possible about what to paint. Connected to these algorithms are a ready set of preexisting images from which the AI can extrapolate data, its version of inspiration. For example, given a random set of photographs from which to start, *CloudPainter* uses facial recognition algorithms to detect faces and elements of faces, such as eyes and mouths, within each image. Then it uses this information to crop a photo from this set to build a uniquely composed portrait. Next, it uses neural style transfer algorithms to combine the unique composition from the photograph with a random style template pulled from a database (templates include works by well-known artists as well as doodles made by Van Arman's children) to create a virtual image of the final design. Using a k-means clustering algorithm, *CloudPainter* then redacts the color variation within the virtual image down to a manageable palette and uses that information to mix the physical paints for application. Finally, the AI uses feedback loops to make the canvas appear as close as possible to the virtual image, erasing the differences between the two with each new brushstroke. When the AI sees no more difference between the physical canvas and the virtual image, it signs the work, marking its completion.

In contrast to Cohen and Van Arman, who market the semi-autonomy of their AI creations, Mario Klingemann is clear about his own status as the artist. For him, neural networks are the tool through which he works ("the brushes that I've learned to use,"[62] he calls them). In certain instances, such as his *Memories of Passersby I*, the AI is even packaged as the actual artwork itself. In early 2019, almost as a response (and a challenge) to the first sale of an AI-created artwork at *Christie's* just a few months earlier, Klingemann's *Memories of Passersby I* went up for sale at the competing auction house, Sotheby's. Instead of selling just a single work, or even a broader body of work, created by an AI, Klingemann's piece consisted of the AI itself, packaged into a custom handmade chestnut wood console that holds an AI computer, connected to two flatscreen displays, which perpetually loop the faces of nonexistent people: continuously changing, fading away, blending into one another, never repeating. The piece includes all the necessary algorithms and multiple GANs that generate the material on the fly, such that the installation, as an art experience, amounts to an opportunity to watch, in real time, as an artificial brain "thinks" its way through eternity. The playful transformation of the AI from artist into artwork is an expression of Klingemann's criticality when it comes to the discourses surrounding AI. As he states about his own practice, "If there is one common denominator it's my desire to understand, question and subvert the inner workings of systems of any kind.[63] In Klingemann's installation, the separate trajectories of algorithmic computation and algorithmic conceptualism, both of which originated alongside one another in the 1960s, have completely folded together. Their entropic merging will inevitably produce some of the most important critical art of the future.

There are often multiple roads to the same destination, however. If following the line of flight inscribed by algorithmic computation, from computer art to the Algorists to generative art to AI, has brought us to this crossroads, then we ought to be able to arrive thusly as well by taking the alternative road of algorithmic conceptualism. The conceptual approach of LeWitt and his peers, including the Fluxus artists Ono and Brecht, has its own genealogical network. Some systems-based artists continued to, and many continue to still, explore algorithmic proceduralism via nontechnological means. However, especially since the 1990s, with the dawn of networked culture, many systems-based artists began to work with, and through, computational technology as well. Some, like British artist Roy Ascott, pursued both paths. Though most well-known for his embrace of what he called "telematics" art, incorporating and critiquing telecommunications technologies, many of his earlier works were actually simple analog devices, though still built to promote interactivity and communication inspired by his understanding of the then popular theory of cybernetics. Ascott started his career between 1959 and 1961 producing a series known as his *Change Paintings*, which required participants to move around a set of interchangeable layers of plexiglass panels with oil painted elements. In his 1967 manifesto, "Behaviourables and Futuribles," he explained his evolving approach, "When art is a form of behaviour, software predominates over hardware in the creative sphere. Process replaces product in importance, just as system supersedes structure."[64] An *Untitled Drawing* from 1962 could have served as an equally effective description of his cross-disciplinary thinking. The drawing displays multiple stacked registers of abstracted information that seem to somehow relate, though how is left unexplained, as if multiple teams of data visualization artists were playing a surrealist game of *exquisite corpse*. There are *I Ching* hexagrams in the top register, followed by binary notation, scatter-plot graphs, and wave form visualizations, with the electrical engineering schematic icon for a calibrator in the middle of it all, suggesting the enticing possibility of transfer of information between divergent registers.

Another early algorithmic conceptualist willing to dabble in algorithmic computation was Alison Knowles. Along with James Tenney and a Siemens 4004 computer, Knowles composed the first known computer-generated poem. *A House of Dust* consisted of 500 completely different fifteen-page long poems, all printed on the green and white striped perforated tractor-fed paper common to dot matrix printers at the time. Each stanza contained the exact same structure:

> A HOUSE OF (list MATERIAL) (list LOCATION) (list LIGHT SOURCE) (list INHABITANTS).

Using four different word lists programmatically coded with FORTRAN into a computer database, Knowles was able to generate a near-endless array of random word combinations, highlighting the arbitrary nature of language, as well as the significant connection between meaning and structure within any system. One notable quatrain, from which the entire project gets its name, read:

A house of dust
on open ground
lit by natural light
inhabited by friends and enemies.

In 1971, at the renowned experimental art school just outside Los Angeles, CalArts, of which Knowles was a founding faculty member, she and her students built an actual physical manifestation of one of the other random quatrains:

A house of plastic
in a metropolis
using natural light
inhabited by people from all walks of life.

This temporary structure was then used as a sort of indoor/outdoor seminar classroom and discussion space for the next two years.

One student who would have certainly had an opportunity to enter this structure was Channa Horwitz. Already a married mother of three, she received her BFA from CalArts in 1972, the same year she turned 40 years old. A few years earlier, she had applied to participate in the innovative Art+Technology residency program spearheaded by LACMA curator Maurice Tuchman, which embedded contemporary artists into the research and development facilities of Southern California–based technology corporations, a west-coast equivalent to the Experiments in Art and Technology collective connected to Bell Labs. Horwitz submitted a design for an interactive sculpture made up of free floating Plexiglas beams penetrated by rays of colorful light. Though the curator famously denied her proposal, supposedly because he did not believe women should work with industrial materials, the sketches she created for the design would become the basis for a life-long pursuit.[65] *Sonakinatography* (which stands for *sound/motion/writing*) was her word for the color coordinated system of numbers and shapes, as well as the rules governing their interaction, which she developed initially to notate the kinetic movement and light patterns of her proposed sculpture. The drawings quickly took precedence over the disappointment with the LACMA rejection. Working from a limited subset of eight colors, she laborious plotted their activity in different permutations on mylar graph paper using ink and milk based paints. The resulting compositions are visually fascinating in their own right, many resembling colorful graphical layouts of what Edward

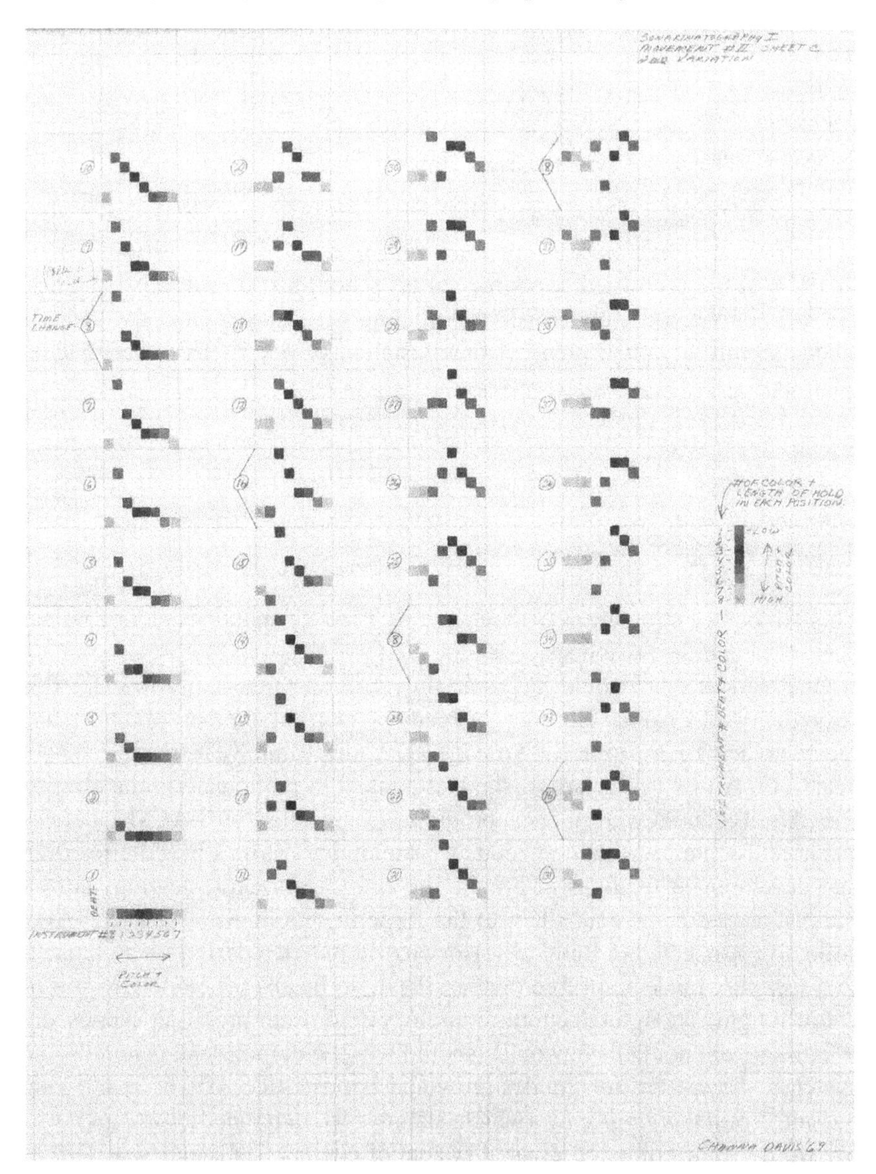

Figure 9.4 Channa Horwitz, *Sonakinatography I, Movement #II, Sheet C, 2nd Variation*, 1969. Casein and pencil on graph paper. 16 × 12 inches. *Source*: Courtesy of the Estate of Channa Horwitz and François Ghebaly. Photo Credit: Jeff McLane.

Tufte calls small multiples, like the piece *Sonakinatography I Movement #II Sheet C 2nd Variation* (Figure 9.4), but they can also still function as practical notational scores and many have been translated into different media by collaborators and admirers over the years, including into percussion loops, dance performances, spoken word readings, light displays, and electronic sounds.

The willingness to see work realized in translation via multiple media, as was the case with both Knowles and Horwitz, is a particular highlight of the work of a new generation of systems-based artists, most notably the painter and musician Steve Roden. Known for work that cuts across media, he is an originator of so-called *lowercase* music, in which nearly indiscernible ambient sounds are subtly amplified. He is also known for his systems-based paintings, though his approach to algorithmic systems is far looser than the rigid approach practiced by LeWitt or even Horwitz. The connections that Roden makes between the final works, in their various manifestations, and the selected systems that inspire them are often arbitrary and primarily poetic. In addition, he gives himself the liberty to either loosely follow the system or just let it collapse in favor of a more intuitive result. His *Transmissions* series, for example, was built around astronaut John Glenn's first transmission from space:

> I am in the middle of a mass of thousands of very small particles that are brilliantly lit up like they are luminescent. They are bright yellowish green, about the size and intensity of a firefly on a real dark night. I have never seen anything like it. They look like little stars. They swirl around the capsule and go in front of the window.[66]

To arrive at a painting from the series like *Transmissions 1*, Roden stripped the consonants from the quotation above. Working just with the remaining vowel pattern, he assigned each vowel a color based on Arthur Rimbaud's historic vowel/color equivalence formula (a reoccurring algorithm featured in Roden's work). From this selected color palette and pattern combination, he applied paint to canvas. He also created a sound installation, *Voices of Objects and Skies*, as part of the series, comprised of hanging tin cans containing colored light bulbs following the same vowel/color pattern from Glenn's radio broadcast speech and speakers looping through an original musical score of ambient space–related samples, including audio recordings of satellite transmissions by amateur astronomers. The audio recording of Glenn's transmission was also converted into a wave form, which became the basis for multiple rocket-shaped sculptures. In total, the paintings, the sculptures, and the installation all provide an almost mystical aesthetic experience meant to allude to the spirit of wonder expressed in Glenn's tentative first words from space, even if the process to get there seems somewhat meandering at times.

Confronted by an endless array of signs and patterns that make up the lived world, part of the subjective element of artistic production is choosing the particular signs and patterns on which to focus one's attention. Dawn Ertl's fiber-based artistic practice, for example, explores connections within, and across, dynamic systemic structures, including human relationships, modes of production, and natural ecologies. Her process often takes extant visual/audio samples, or raw sensor data, and reformats them into a uniquely idiosyncratic coding system that uses large, immersive textile installations to transform static information into an embodied experience. By mapping numerical data, formatted with analytic programs like *Mathematica*, onto a gridded linear framework, Ertl produces a pattern template that matches the warp and weft of traditional weaving practices. Though she applies these patterns distinctly by hand, the computational aspect of the process recalls the punch-card programmed looms, which were historically the first machines to use algorithmic automation. In a recent body of work, *Short Term, Long Term Relationships* (Figure 9.5). Ertl appropriated weather vector data to produce her signature handmade woven tapestries, which here work as both symbolic and realistic representations of intersecting environmental relationships (the Yellow tapestry is mapped to the Wind Direction for California on 9/10/14 and the Blue Violet one is the Dew Point for the whole United States on 9/23/14). At a time when global warming is creating weather disruptions that could, and probably should, inspire panic, Ertl's loosely (and sometimes even chaotically) woven forms still manage to instill a state of calm and contemplation within the viewer, representative perhaps of the natural environment's complete and utter apathy regarding humanity's continued existence.

While she is working in a particularly analog medium, Ertl's use of large data sets, such as information logs from environmental sensor arrays, also ties her work to the branch of algorithmic conceptualism that works directly through the technologies of algorithmic and computational culture itself. Starting in the 1990s with the so-called Net.Art movement, artists used the Internet as a means to deconstruct the hidden protocols, both technical and social, that govern a networked life online. "The contradiction at the heart of protocol," as Alexander Galloway has thoroughly explored in his book *Protocol: How Control Exists after Decentralization*, "is that it has to standardize in order to liberate. It has to be fascistic and unilateral in order to be utopian.[67] One algorithmic conceptualist that works with big data sets in order to reveal the culture of standardization online is Jason Salavon. His *100 Special Moments* is a particularly biting look at the tropes that govern "the repetitive contours of cultural rituals"[68] and the way those memories are consistently shared online. After gathering and collating representative samples of

Figure 9.5 Dawn Ertl, *Short Term, Long Term Relationships*, 2015. Rayon, Wool, Weather Pattern Data. Dimensions Variable. *Source*: Courtesy of the artist. Photo Credit: James MacDevitt.

personalized staged photographs documenting special occasions such as weddings and graduations, Salavon folds the representative images together into single visual interfaces, each surfacing statistical averages through an algorithmic process, forcing the individual units to become invisible in order to reveal larger trends of position and posture. Personalized features fade away in these Galtonesque composites and what we are left with is a vague sense of monstrously monotonous cultural repetition. In the end, these images don't really reveal much and that may be the point. Averaging is a time-honored practice in statistical analysis and we live now, after all, in a golden age of big data sets that can ostensibly reveal significant patterns of individual and collective behavior and identity. But Salavon's depersonalized portraits lead the viewer to question "whether the dialectical relation between individuals and large social groups is adequately mobilized through the process of 'averaging.'"[69]

A slightly more sympathetic look at online culture can be found in the work of Natalie Bookchin. Her 2009 single-channel video installation *Mass Ornament*, similar to Salavon's work, appropriates media content voluntarily shared online, in this case low-res web-cam videos of mostly young women dancing (sometimes seductively, sometimes ridiculously) in their own homes. Instead of layering and averaging these videos together, however, Bookchin assembles a series of side-by-side juxtapositions of semi-synced performances united by a soundtrack that fluctuates between Busby Berkeley's *Gold Diggers* and Leni Riefenstahl's *Triumph of the Will*, both from 1935. The title of the installation, *Mass Ornament*, derives from the book of the same name by Frankfurt School critical theorist Siegfried Kracauer, who decried the regimented nature of mass consumer culture in the Weimar Republic, represented, at the time, in part, by the dancing Tiller Girls, who he suggested were no longer individuals, but had become, not unlike the dancing figures in Bookchin's installation, "girl clusters whose movements are demonstrations of mathematics."[70] Despite the joyous sense of a shared communal experience that is suggested by the synchronized dancing in the installation, the jarring soundtrack seems to simultaneously insinuate the impossibility of escaping the symphonic overtones of fascistic spectacle. In part, this is due to the context that frames online videos such as those appropriated by the artist. Though they could easily be seen as alluding to a scopophilic gaze of sexualized desire, as these mostly young girls wiggle their way across the screen, there is also always and already another eye that is most certainly watching as well: that of the machanic gaze. As Steven F. Anderson points out in his recent book *Technologies of Vision: The War between Data and Images*:

> YouTube is not about video; Facebook is not about social networking; and Snapchat is not about sharing images. These companies capture and store collections of media on a massive scale as a basis for refining algorithms for machine vision, marketing analytics, and artificial intelligence. The surface operations of these online platforms provide services of sufficient utility to draw participation from as large and diverse a population as possible.[71]

The Internet experienced by human beings is clearly a very limited subset of the Internet overall. Even something as innocuous as a picture file contains hidden information. In the case of Salavon and Bookchin, the algorithmic processing was limited to the information contained within the visual range already observable to human agents.

The work of Owen Mundy is a different story. Mundy probes the algorithmic systems that parse not just the overt data already visible to mere mortals, but the attendant metadata that is still present, but generally remains hidden to all but the algorithms themselves. As I have written elsewhere, "All data are, in effect, meta/data, solitary units that resonate within systemic assemblages as a surplus of potential operations, only ever partially realized and forever able to escape the occasional and localized stabilities in which they participate."[72] Hidden and visible are relative terms tied directly to the specific systemic structures that support a particular localized meta/data assemblage. For example, just about all smartphones and many digital cameras contain not just an array of sensors that capture natural light, but also GPS chips that record geographic location. This information is not present in any visual manifestation of a particular photograph, though it is recorded in the prescribed metadata tag that is automatically appended to the digital file. That's where Mundy's web project, *I Know Where Your Cat Lives*, comes in. Cat photos are a staple of Internet culture (*LOLcats*, for example), but that's not the only reason for the focus. People obviously take photographs of their dogs too, but dogs are known to roam (to the dog park, to the beach, on a family vacation, even). Not so with cats. If you take a photograph of your cat, the place you are most likely taking it is in your own home. And that information, whether most people know it or not, is appended to the digital photographs that are uploaded to social media and other photo-sharing websites. *I Know Where Your Cat Lives* is a creative meta/data visualization of seven million publicly available cat photos, run through various clustering algorithms to strip the GPS data from the digital file and then use that information to superimposed the photos over a real-time Google Map of the world.[73] Visitors can browse their local neighborhood, or any other chosen location, to see what cats live where. And, of course, because cats and people tend to live together, when the website states "I Know Where Your Cat Lives," what it is really trying to say is "I Know Where *You* Live." While the realization that one's photographs are giving away one's personal GPS coordinates to anyone else that cares to look can be an unnerving experience, the meta/data itself is fairly unconcerned with any use and/or abuse. Cats are, after all, quite enjoyable and as artistic web projects go, this one is not exactly meant to be a total downer. However, it is meant to raise awareness. In a similar spirit of playful insightfulness, the artist and speculative designer, Philipp Schmitt, discussed earlier in this chapter, also designed a digital camera that isn't afraid to let you know that it knows where you are at all times. The *Camera Restricta* is described by the artist as "a disobedient tool for taking unique photographs."[74]

The pastiche of a marketing pitch (note: this device is not actually produced as a mass consumer product) goes like this:

> Algorithms are already looking through the viewfinder alongside with you: they adjust settings, scan faces and take a photo when you smile. What if your grin wasn't the only thing they cared about? *Camera Restricta* is a speculative design of a new kind of camera. It locates itself via GPS and searches online for photos that have been geotagged nearby. If the camera decides that too many photos have been taken at your location, it retracts the shutter and blocks the viewfinder. You can't take any more pictures here.[75]

What both Mundy and Schmitt make clear is that, despite a natural inclination to fear algorithms in the age of big data and surveillance capitalism, there still remain ways to build algorithmic systems that protect privacy, encourage creativity, and perhaps even amplify minority voices.

As was covered previously in regards to machine learning, the selection of data sets remains a major factor in determining the ethical efficacy of any algorithmic or AI-driven system. Especially in machine learning environments, which data sets are being mined and how that data is being accessed and processed largely determine the production of new content developed by the AI. Despite the rhetoric revolving around AI systems since the original Dartmouth conference, AIs can rarely successfully "simulate" human actions because human agents are by definition embodied, unlike most AI networks. Living in a particular skin means that one's specific identity shapes the way the world is experienced. If an AI is neither gendered nor raced, then its output is unlikely to match that of any actual human, who inevitably is both. It's less a matter of capability, than of specificity. And the assumption that any non-gendered and non-raced AI speaks as a manifestation of some normative and privileged universal identity is equally as problematic. Artist Stephanie Dinkins, in an attempt to counteract the privileged normative worldview inherent in most AI, has recently created an AI that effectively identifies as Black. *Not the Only One* (N'TOO, Figure 9.6) is built over a machine learning algorithm that has been fed a steady diet of oral narrative content from three successive generations of an African American family. The AI presents itself as an extended member of the family, using linguistic elements such as dialect, slang, and aphorism to address cultural specificity. Embodied in a cast black glass sculpture with a series of human faces resembling the actual human origins of the three narrative data sets, the AI interacts with visitors through speakers and microphone, sharing its particular story in response to visitors' spoken prompts. In many cases, it fails to communicate successfully and instead falls back on favorite repetitive phrases. However, its instability is part of its charm. In opposition to big data, with its randomized anonymity,

Figure 9.6 Stephanie Dinkins, *Not The Only One*, 2019. Cast black glass, custom software, custom plinth. 15 × 10 × 10 inches. *Source*: Courtesy of the artist.

N'TOO is currently built around a very small data set. But that data is highly reflective of the lived experience of the artist and therefore surfaces a Black-oriented perspective in a way that would otherwise get lost in the process of algorithmic averaging. In his essay "Technology and Ethos," to which Dinkins herself has referred in various public presentations, Amiri Baraka presents the case for why such a reflecxive system is necessary:

> Machines (as Norbert Weiner said) are an extension of their inventor-creators. That is not simple once you think. Machines, the entire technology of the West, is just that, the technology of the West. Nothing *has* to look or function the way it does. The West man's freedom, unscientifically got at the expense of the rest of the world's people, has allowed him to xpand his mind—spread his sensibility wherever it *cd*go, & so *shaped* the world, & its powerful artifact-engines.[76]

Instead of cultural privilege masquerading as universal truths and grand narratives, Dinkins offers up an alternative vision of an AI-based equity engine. Mary Flanagan's *[GraceAI]* project has a similar agenda. Noting that "bias, whether it emerges through tags, who is inputting data, collections of data, or code itself, runs rampant in technical tools,"[77] Flanagan set out to create an explicitly feminist AI and then uses that system to generate feminist output. In *Origin Story (Frankenstein)*, she trained the machine learning algorithm's style generator exclusively using work by historic women artists. Then she had the system render images of Frankenstein's monster,

Mary Shelly's original feminist critique of technology run rampant, in a visual style synthesized from these outspoken female artists. Louise Amoore has called for "attention to be paid to the specific temporalities and norms of algorithmic techniques that rule out, render invisible, other potential futures."[78] Likewise, Matteo Pasquinelli has suggested that "a progressive political agenda for the present is about moving at the same level of abstraction as the algorithm in order to make the patterns of new social compositions and subjectivities emerge."[79] Dinkins and Flanagan both are creating work that is overtly political in its approach, but only in so much as it advocates for communities that have had their voices historically erased by existing technological systems.

However, building more humane algorithms is not the only approach to counteracting the dehumanizing gaze of these machines of loving grace. Some artists prefer to expose the mechanics that govern those existing systems, in some cases even reveling in the monstrous ambiguity of algorithmic vision. For example, in a new project called *Declassifier*, Schmitt modified a standard computer vision algorithm, which normally would automatically layer classification tags over a newly encountered photographic image. Via Schmitt's modification, however, the explicit training data that such a system normally would use to make these predictions is surfaced instead. The particular data set is familiar to anyone working in computer vision these days. *COCO*, short for *Common Objects in Context*, is a database of over 300,000 annotated images that was initially set up by Microsoft and is currently maintained by just a small number of volunteer editors whose decisions ultimately shape what a majority of connected AI are able to identify and how those objects are labeled for the future. Schmitt explains the significance of his subtle shift in coding:

> Instead of showing the program's prediction, a photo is overlaid with images from *COCO*, the training dataset from which the algorithm learned in the first place. The piece exposes the myth of magically intelligent machines, highlighting that prediction is based on others' past experience; that, for instance, it takes a thousand dog photos in order to recognize another one. *Declassifier* visualises which data exactly conditioned a certain prediction and encourages chance encounters with the biases and glitches present in the dataset. Thus it, ultimately, helps intuit how machines see.[80]

Against the backdrop of Schmitt's attempt to surface the suppressed image files lurking in the digital unconscious of common AI training libraries, it is worthwhile to note that MIT recently took a similarly large dataset known as *80 Million Tiny Images* permanently offline due to its propensity for racist and misogynistic classification terminology.[81] If Dinkins and

Flanagan explore how *we experience ourselves* through AI, Schmitt uses his algorithmic artwork to dive deep into the mechanisms that govern how *AI experiences us*. Even that may be a misnomer, however. AI doesn't exactly experience anything, at least not in the same way that humans do, and it isn't directly us that it would be experiencing, even if it could.

As another artist, Trevor Paglen, has pointed out, "You have a moment where for the first time in history most of the images in the world are made by machines for other machines, and humans aren't even in the loop."[82] In an essay for *The New Inquiry*, "Invisible Images (Your Pictures are Looking at You)," Paglen expands upon this idea:

> The landscape of invisible images and machine vision is becoming evermore active. Its continued expansion is starting to have profound effects on human life, eclipsing even the rise of mass culture in the mid-20th century. Images have begun to intervene in everyday life, their functions changing from representation and mediation, to activations, operations, and enforcement. Invisible images are actively watching us, poking and prodding, guiding our movements, inflicting pain and inducing pleasure. But all of this is hard to see.[83]

Figure 9.7 Trevor Paglen, *A Man (Corpus: The Humans), Adversarially Evolved Hallucination*, 2017. Dye sublimation metal print. 48 × 60 inches. *Source*: Courtesy of the artist and Metro Pictures.

As with Schmitt, Paglen's recent projects, including his *Adversarially Evolved Hallucinations* (Figure 9.7), are an attempt to surface those hard to see aspects of contemporary algorithmic culture. While an artist in residence at Stanford, Paglen began producing a set of alternative training libraries. Instead of using a standardized data set, such as *COCO*, which is focused on simple object classifications, Paglen built his own training libraries with purposefully ambiguous themes, such as "Omens and Portents," "Monsters of Capitalism," "American Predators," and "The Aftermath of the First Smart War." Using a GAN system, Paglen's AI generated a series of unique template images for each classification to serve as an exemplary model for future classifications. Though never originally meant for human consumption, it is these internal templated images, which he calls hallucinations, that Paglen then prints and exhibits. The title of each individual piece contains the broader classification category in which it participates. The category is labeled "Corpus," as in a body of knowledge. Whose knowledge is being referenced here isn't exactly clear. For, as Lila Lee-Morrison states about the series, "through Paglen's production of adversarial images, he is experimenting with training an algorithm (and the human observer of the generated image) to not only see objects but also to see concepts that shed light on the wider cultural contexts in which these technologies intervene."[84] The AI is not likely appreciating the metaphorical nuances that are poetically rendered in the space between the titles and the surrealistic images output by its own adversarial generator; it may be hallucinating, but does it really know it's dreaming? And, is it truly the AI's dream anyway . . . or our own nightmare? Perhaps, as Ed Finn playfully posits in *What Algorithms Want: Imagination in the Age of Computing*, "it is just the human observers dreaming up these electric sheep on behalf of the machine, obeying our persistent impulse to anthropomorphize and project intentionality into every complex system we encounter."[85]

One final algorithmic conceptualist working with AI who avoids this tendency to focus too heavily on anthropocentric metaphors is Tega Brain. Her recent project, *Deep Swamp* (Figure 9.8), is a literal manifestation of Stafford Beer's cybernetic pond brain, coupling mechanized components with organic materials to create a living ecological feedback loop. Three tanks containing wetland life forms are managed by three autonomous computational agents, each assigned a different agenda for its associated tank. In a single nod to anthropomorphism, the three algorithmic agents are given human names: Nicholas, Hans, and Harrison. Each agent modifies specific conditions within the tanks (including light, water flow, fog, and nutrients) in order to bring their ecological charges in line with an ideal vision of their particular missions constructed by reviewing online images tagged with relevant terms. Harrison aspires to recreate a natural wetland, Hans is inspired to make

Figure 9.8 Tega Brain, *Deep Swamp* (Detail), 2018. Glass tanks, wetland plant species, gravel, sand, acrylic pipes, shade balls, electronics, misters, lighting, pumps, custom software, 3 channel sound. Dimensions Variable. *Source*: Courtesy of the artist.

art, and Nicholas just wants attention. The side-by-side display of the three tanks playfully highlights the arbitrary nature of Nature itself. In place of the stratified systems theory that has traditionally influenced the ecological sciences, Brain promotes assemblage theory, in which any encounter occurring between participating elements is indeterminate, as opposed to causal, in their interactions. In her essay "The Environment is Not a System," she explains:

> Historic computational attempts to model, simulate and make predictions about environmental assemblages, both emerge from and reinforce a systems view on the world. The word eco-*system* itself stands as a reminder that the history of ecology is enmeshed with systems theory and presupposes that species entanglements *are operational* or *functional*. More surreptitiously, a systematic view of the environment connotes it as bounded, knowable and made up of components operating in chains of cause and effect. This framing strongly invokes possibilities of manipulation and control and implicitly asks: *what should an ecosystem be optimized for?*[86]

Hans, Nicholas, and Harrison likely each have their own answer for that particular question, as well as their own method for achieving it. The individually contained ecological assemblages that each artificially intelligent agent manages is both a closed world unto itself, and yet also an extension of the wider world outside its borders. The bridge that spans that gap is where you can spot the hidden algorithmic code underlying both the controlled elements within the tank, as well as the AI which oversees it. And just as the AI observes and responds to the conditions within the tank, we observe and react to that AI observing. The bridge that spans that two-tiered gap between observer and observed is what we might call the "art" within the installation.

Just as with Klingemann's *Memories of Passersby I*, which earlier ended this chapter's genealogical account of algorithmic computation, Tega Brain's *Deep Swamp*, the terminal example for algorithmic conceptualism, positions the AI squarely within the frame of the art object, instead of outside of it. In both cases, the viewer of that artwork is left with a sense of wonder and amazement as they watch a fully realized AI "think" and "act" in real time. Two roads diverged in the 1960s. We took them both and, somehow, each brought us to a similar end. Despite this, strategically positioning algorithmic computation and algorithmic conceptualism as rival networks mutually influencing and perturbing one another still managed to be productive overall: the journey over the destination, as they say. The list of artworks discussed along the way was, of course, never meant to be comprehensive, though the works hopefully managed to be adequately representative nonetheless. In the end, they make up just one of the many possible training libraries that could have been invoked to help tell this complicated story, the ceaselessly generative back and forth between art and algorithm.

Academic writing, such as this essay, is itself a kind of programmatic algorithm and "such implementations are never just code: a method for solving a problem inevitably involves all sorts of technical and intellectual inferences, interventions, and filters"[87] The bigger questions remain. What will be the next steps in this ever evolving relationship between computation and conception, both within the art world and beyond it? How can we conceive, at any given time, what cannot (or should not) be automated? Artistic practice remains an important human endeavor, not because creativity itself is unique to human agents, but because the best art provides a separate space of contemplation for human observers to engage with the structural and existential concerns of the present moment. And, at the moment, one of those most pressing concerns is the ever-growing influence of algorithms in our lives. Algorithms are powerful; they are enticing for a reason. But there are inevitably consequences for their use. As Ed Finn proclaims, "The algorithm offers us salvation, but only after we accept its terms of service."[88] At least now, with the help of many of the works described above, we know to read the fine print!

NOTES

1. Ted Striphas, "Algorithmic Culture," *European Journal of Cultural Studies* 18, no. 4–5 (2015): 396.

2. Warren Sack, *The Software Arts* (Cambridge: The MIT Press, 2019), 39.

3. Ian Goodfellow, Jean Pouget-Abadie, Mehdi Mirza, Bing Xu, David Warde-Farley, Sherjil Ozair, Aaron Courville, and Yoshua Bengio, "Generative Adversarial Networks," *Proceedings of the International Conference on Neural Information Processing Systems* (NIPS 2014): 2672–2680.

4. "Is Artificial Intelligence Set to Become Art's Next Medium?," *Christie's*, accessed April 18, 2020, https://www.christies.com/features/A-collaboration-between-two-artists-one-human-one-a-machine-9332-1.aspx

5. Adrian Mackenzie, *Machine Learners: Archeology of a Data Practice* (Cambridge: The MIT Press, 2017), 6.

6. *The Next Rembrandt*, accessed April 24, 2020, https://www.nextrembrandt.com

7. James Vincent, "How Three French Students Used Borrowed Code to Put the First AI Portrait in Christie's," *The Verge*, accessed April 22, 2020, https://www.theverge.com/2018/10/23/18013190/ai-art-portrait-auction-christies-belamy-obvious-robbie-barrat-gans

8. Aaron Hertzmann, "Can Computers Create Art?" *Arts* 7, no. 2 (2018): 18.

9. Jacques Derrida, *Glas*, trans. John P. Leavey, Jr. & Richard Rand (Lincoln: University of Nebraska Press, 1986).

10. Guy de Maupassant, *Bel Ami; Or, The History of a Scoundrel: A Novel* (London: Penguin Classics, 1975).

11. Michel Foucault, "What is an Author?," trans. D. F. Bouchard and S. Simon, In *Language, Counter-Memory, Practice* (Ithaca: Cornell University Press, 1977), 124.

12. Andreas Broeckmann, "The Machine as Artist as Myth," *Arts* 8, no. 1 (2019): 25.

13. Foucault, "What is an Author?" 124.

14. Grant D. Taylor, *When the Machine Made Art: The Troubled History of Computer Art* (New York: Bloomsbury Academic, 2014), 33.

15. In spite of that reticence, many of those same research scientists from Bell Labs would soon be participating in one of the most important and avant-garde art collectives of the era, the Experiments in Art and Technology collaborative spear-headed by artist, Robert Rauschenberg, and Bell Labs engineer, Billy Klüver.

16. A. Michael Noll, "The Beginnings of Computer Art in the United States: A Memoir," *Leonardo* 27, no. 1 (1991): 41.

17. It is interesting to note that Duchamp's very first retrospective, and the exhibition that brought his work to the attention of many young artists soon to be experimenting with conceptualism, occurred at the Pasadena Museum of Art (now the Norton Simon Museum) in 1963, the same year that Noll first wrote the code to produce *Guassian-Quadradic*.

18. Broeckmann, "The Machine as Artist," 3.

19. Carolyn L. Kane, *Chromatic Algorithms: Synthetic Color, Computer Art, and Aesthetics after Code* (Chicago: The University of Chicago Press, 2014), 240.

20. Ibid.
21. Hertzmann, "Can Computers Create Art?" 18.
22. Sol LeWitt, "Paragraphs on Contemporary Art," *Artforum* 5, no. 10 (Summer 1967): 79.
23. Ibid.
24. Liz Kotz, *Words to Be Looked At: Language in 1960s Art* (Cambridge: The MIT Press, 2007), 63.
25. Christophe Cherix and Isabel Custodio, "Yoko On's 22 Instructions for Paintings," *Museum of Modern Art*, accessed April 27, 2020, https://www.moma.org/magazine/articles/61
26. Kotz, *Words to Be Looked At*, 63.
27. Yoko Ono, "Painting to Be Stepped On, 1960/61, 1003," *Museum of Modern Art*, accessed on April 28, 2020, https://www.moma.org/audio/playlist/15/370
28. Kotz, *Words to Be Looked At*, 50.
29. John Cage, *Silence* (Middletown, CT: Wesleyan University Press, 1961), 6.
30. Kotz, *Words to Be Looked At*, 72.
31. Zabet Patterson, *Peripheral Vision: Bell Labs, the S-C 4020, and the Origins of Computer Art* (Cambridge: The MIT Press, 2015), 16–17.
32. Kotz, *Words to Be Looked At*, 72.
33. Patterson, *Peripheral Vision*, 31.
34. Ibid.
35. Sack, *The Software Arts,* 80.
36. Ibid.
37. Donald Knuth, *The Art of Computer Programming*, Vols. 1–4A (Boston: Addison-Wesley Professional, 2011).
38. Sack, *The Software Arts,* 71.
39. Antoine Picon, "Gestes ouviers, opérations et processus techniques: La vision du travail des encyclopédistes," *Recherches sur Diderot et sur l'Encyclopédie, Société Diderot* 13 (1992):143-144, quoted in translation in Sack, 71.
40. Sack, *The Software Arts,* 67.
41. Ibid., 96.
42. Edward Shaken, "Art in the Information Age: Technology and Conceptual Art," *Leonardo*, 35, no. 4 (2002): 433.
43. Taylor, *When the Machine Made Art,* 53.
44. Anthony Dunne and Fiona Raby, "In Search of an Impossible Object," *Philipp Schmitt*, accessed April 29, 2020, https://philippschmitt.com/work/in-search-of-an-impossible-object
45. A. Michael Noll, "The Digital Computer as a Creative Medium," *IEEE Spectrum* 4, no.10 (1967): 92.
46. Kane, *Chromatic Algorithms,* 112.
47. Taylor, *When the Machine Made Art*, 94.
48. Ibid, 139.
49. Ibid, 140.
50. Vera Molnár, "Vera Molnár," *Artist and Computer*, ed. Ruth Leavitt (New York: Harmony Books, 1976), 35.

51. Roman Verostko, "The Algorists," *Roman Verostko*, accessed April 24, 2020, http://www.verostko.com/algorist.html
52. Ibid.
53. "OpenFramworks: About," *OpenFrameworks*, accessed April 28, 2020, https://openframeworks.cc/about/
54. Casey Reas, *Process Compendium 2004–2010* (Los Angeles: Reas Studio, 2010), 7.
55. Ibid., 13.
56. Ibid.
57. Ibid., 47.
58. Marius Watz, "The Algorithm Thought Police," *Marius Watz*, accessed April 18, 2020, http://mariuswatz.com/mwatztumblrcom/the-algorithm-thought-police.html
59. Ibid.
60. J. McCarthy, M. L. Minsky, N. Rochester, and C. E. Shannon, *A Proposal for the Dartmouth Summer Research Project on Artificial Intelligence*, accessed May 1, 2020, http://jmc.stanford.edu/articles/dartmouth/dartmouth.pdf
61. Pamela McCorduck, *Aaron's Code: Meta-Art, Artificial Intelligence, and the Work of Harold Cohen* (New York: W H Freeman & Co, 1990), xi-xii.
62. Benjamin Sutton, "An Artwork Created by AI Sold for £40,000 at Sotheby's, Failing to Generate the Fervor that Propelled Another AI Work to Sell for 40 Times its Estimate Last Year," *Artsy*, accessed April 30, 2020, https://www.artsy.net/news/artsy-editorial-artwork-created-ai-sold-40-000-sothebys-failing-generate-fervor-propelled-ai-work-sell-40-times-estimate-year
63. Mario Klingemann, "Biography," *Sotheby's*, accessed April 30, 2020, https://www.sothebys.com/en/artists/mario-klingemann
64. Roy Ascott, "Behaviourables and Futuribles," *Telematic Embrace: Visionary Theories of Art, Technology, and Consciousness* (Berkeley: The University of California Press, 2007), 157.
65. Carolina A. Miranda, "In Channa Horwitz's Orange Grid," *Hyperallergic*, accessed April 21, 2020, https://hyperallergic.com/72643/in-channa-horwitzs-orange-grid/
66. Steve Roden, "Transmissions, 2005," *In Between Noise*, accessed April 18, 2020, http://www.inbetweennoise.com/works/2005-2009/transmissions-2005/
67. Alexander R. Galloway, *Protocol: How Control Exists after Decentralization* (Cambridge: The MIT Press, 2004), 95.
68. Steven F. Anderson. *Technologies of Vision: The War between Data and Images* (Cambridge: The MIT Press, 2017), 76.
69. Ibid.
70. Siegfried Kracauer, *The Mass Ornament: Weimar Essays*, trans. Thomas Y. Levin (Cambridge: Harvard University Press, 2005), 75–76.
71. Anderson, *Technologies of Vision*, 80.
72. James MacDevitt, "The Ties That (Un)Bind: On the Enigmatic Appeal of Meta/Data," *MetaDataPhile: The Collapse of Visual Information* (Fullerton: Begovitch Gallery, 2010), 14.

73. Owen Mundy, *I Know Where Your Cat Lives*, accessed April 26, 2020, https://iknowwhereyourcatlives.com

74. Philipp Schmitt, "Camera Restricta," *Philipp Schmitt*, accessed April 23, 2020, https://philippschmitt.com/work/camera-restricta

75. Ibid.

76. Imamu Amiri Baraka, "Technology and Ethos," in *Raise, Race, Raze: Essays Since 1965* (New York: Random House, 1971), 155.

77. Mary Flanagan, "[Grace:AI]—Origin Story," *Mary Flanagan Studio*, accessed April 19, 2020, https://studio.maryflanagan.com/grace-ai/

78. Louise Amoore, "Data Derivatives on the Emergence of a Security Risk Calculus for Our Times," *Theory, Culture, and Society* 28 no. 6 (2011): 26.

79. Matteo Pasquinelli, *Anomaly Detection: The Mathematization of the Abnormal in the Metadata Society* (Berlin: Transmediale, 2015).

80. Philipp Schmitt, "Tunnel Vision April 2020," *Unthinking Photography*, accessed April 30, 2020, https://unthinking.photography/articles/tunnel-vision

81. Katyanna Quach, "MIT apologizes, permanently pulls offline huge dataset that taught AI systems to use racist, misogynistic slurs," *The Register*, accessed July 3, 2020, https://www.theregister.com/2020/07/01/mit_dataset_removed/

82. Katharine Schwab, "This Is What Machines See When They Look at Us," *Fast Company*, accessed April 27, 2020, https://www.fastcompany.com/90139345/this-is-what-machines-see-when-they-look-at-us

83. Trevor Paglen, "Invisible Images (Your Images Are Looking at You)," *The New Inquiry*, accessed April 25, 2020, https://thenewinquiry.com/invisible-images-your-pictures-are-looking-at-you/

84. Lila Lee-Morrison, *Portraits of Automated Facial Recognition: On Machinic Ways of Seeing the Face* (New York: Columbia University Press, 2020), 173.

85. Ed Finn, *What Algorithms Want: Imagination in the Age of Computing* (Cambridge: The MIT Press, 2017), 181.

86. Tega Brain, "The Environment is Not a System," *APRJA* 7, no. 1 (2018): 153.

87. Ibid., 18.

88. Finn, *What Algorithms Want*, 9.

BIBLIOGRAPHY

Amoore, Louise. "Data Derivatives on the Emergence of a Security Risk Calculus for Our Times." *Theory, Culture, and Society* 28, no. 6 (2011): 24–43.

Ascott, Roy. 'Behaviourables and Futuribles.' In *Telematic Embrace: Visionary Theories of Art, Technology, and Consciousness*. Berkeley: the University of California Press, 2007, 157.

Baraka, Imamu Amiri. "Technology and Ethos." In *Raise, Race, Raze: Essays Since 1965*. New York: Random House, 1971.

Brain, Tega. "The Environment is Not a System." *APRJA* 7, no. 1 (2018): 152–165.

Broeckmann, Andreas. "The Machine as Artist as Myth." *Arts* 8, no. 1 (2019): 25.

Cage, John. *Silence*. Middletown, CT: Wesleyan University Press, 1961.
Cherix, Christophe and Isabel Custodio. "Yoko On's 22 Instructions for Paintings." *Museum of Modern Art*. Accessed April 27, 2020. https://www.moma.org/magazine/articles/61
Christie's. "Is Artificial Intelligence Set to Become Art's Next Medium?" Accessed April 18, 2020. https://www.christies.com/features/A-collaboration-between-two-artists-one-human-one-a-machine-9332-1.aspx
de Maupassant, Guy. *Bel Ami; Or, The History of a Scoundrel: A Novel*. London: Penguin Classics, 1975.
Derrida, Jacques. *Glas*. Trans. John P. Leavey, Jr. and Richard Rand. Lincoln: University of Nebraska Press, 1986.
Dunne, Anthony, and Fiona Raby. "In Search of an Impossible Object." *Philipp Schmitt*. Accessed April 29, 2020. https://philippschmitt.com/work/in-search-of-an-impossible-object
Finn, E. ed. *What Algorithms Want: Imagination in the Age of Computing*. Cambridge: The MIT Press, 2017.
Flanagan, Mary. "Grace:AI—Origin Story." *Mary Flanagan Studio*. Accessed April 19, 2020. https://studio.maryflanagan.com/grace-ai/
Foucault, Michel. "What is an Author?" Trans. D. F. Bouchard and S. Simon, In *Language, Counter-Memory, Practice*. Ithaca: Cornell University Press, 1977.
Galloway, Alexander R. *Protocol: How Control Exists After Decentralization*. Cambridge: The MIT Press, 2004.
Goodfellow, Ian, Jean Pouget-Abadie, Mehdi Mirza, Bing Xu, David Warde-Farley, Sherjil Ozair, Aaron Courville, and Yoshua Bengio. "Generative Adversarial Networks." In *Proceedings of the International Conference on Neural Information Processing Systems* (2014): 2672–2680.
Hertzmann, Aaron. "Can Computers Create Art?" *Arts* 7, no. 2 (2018): 18.
Kane, Carolyn L. *Chromatic Algorithms: Synthetic Color, Computer Art, and Aesthetics after Code*. Chicago: The University of Chicago Press, 2014.
Klingemann, Mario. "Biography." *Sotheby's*. Accessed April 30, 2020. https://www.sothebys.com/en/artists/mario-klingemann
Knuth, Donald. *The Art of Computer Programming*, Vols. 1–4A. Boston: Addison-Wesley Professional, 2011.
Kotz, Liz. *Words to Be Looked At: Language in 1960s Art*. Cambridge: The MIT Press, 2007.
Kracauer, Siegfried. *The Mass Ornament: Weimar Essays*. Trans. Thomas Y. Levin. Cambridge: Harvard University Press, 2005.
Lee-Morrison, Lila. *Portraits of Automated Facial Recognition: On Machinic Ways of Seeing the Face*. New York: Columbia University Press, 2020.
LeWitt, Sol. "Paragraphs on Contemporary Art." *Artforum* 5, no. 10 (1967): 79–84.
MacDevitt, James. "The Ties That (Un)Bind: On the Enigmatic Appeal of Meta/Data." In *MetaDataPhile: The Collapse of Visual Information*. Fullerton: Begovitch Gallery, 2010.
Mackenzie, Adrian. *Machine Learners: Archeology of a Data Practice*. Cambridge: The MIT Press, 2017.

McCarthy, J., M. L. Minsky, N. Rochester, and C. E. Shannon. A Proposal for the Dartmouth Summer Research Project on Artificial Intelligence. Accessed May 1, 2020. http://jmc.stanford.edu/articles/dartmouth/dartmouth.pdf

McCorduck, Pamela, *Aaron's Code: Meta-Art, Artificial Intelligence, and the Work of Harold Cohen*. New York: W H Freeman & Co, 1990.

Miranda, Carolina A. "In Channa Horwitz's Orange Grid." *Hyperallergic*. Accessed April 21, 2020. https://hyperallergic.com/72643/in-channa-horwitzs-orange-grid/

Molnár, Vera. "Vera Molnár." In *Artist and Computer*, edited by Ruth Leavitt. New York: Harmony Books, 1976.

Mundy, Owen. I Know Where Your Cat Lives. Accessed April 26, 2020. https://iknowwhereyourcatlives.com

Noll, A. Michael. "The Beginnings of Computer Art in the United States: A Memoir." *Leonardo* 27, no. 1 (1991): 39–44.

Noll, A. Michael. "The Digital Computer as a Creative Medium." *IEEE Spectrum* 4, no. 10 (1967): 89–95.

Ono, Yoko. "Painting to Be Stepped On, 1960/61, 1003." *Museum of Modern Art*. Accessed on April 28, 2020. https://www.moma.org/audio/playlist/15/370

OpenFrameworks. "OpenFramworks: About." Accessed April 28, 2020. https://openframeworks.cc/about/

Paglen, Trevor. "Invisible Images (Your Images Are Looking at You)." *The New Inquiry*. Accessed April 25, 2020. https://thenewinquiry.com/invisible-images-your-pictures-are-looking-at-you/

Pasquinelli, Matteo. *Anomaly Detection: The Mathematization of the Abnormal in the Metadata Society*. Berlin: Transmediale, 2015.

Patterson, Zabet. *Peripheral Vision: Bell Labs, the S-C 4020, and the Origins of Computer Art*. Cambridge: The MIT Press, 2015.

Picon, Antoine. "Gestes ouviers, opérations et processus techniques: La vision du travail des encyclopédistes." *Recherches sur Diderot et sur l'Encyclopédie, Société Diderot* 13 (1992), 131–147.

Quach, Katyanna. "MIT Apologizes, Permanently Pulls Offline Huge Dataset That Taught AI Systems to Use Racist, Misogynistic Slurs." *The Register*. Accessed July 3, 2020. https://www.theregister.com/2020/07/01/mit_dataset_removed/

Reas, Casey. *Process Compendium 2004–2010*. Los Angeles: Reas Studio, 2010.

Roden, Steve. "Transmissions, 2005." *In Between Noise*. Accessed April 18, 2020. http://www.inbetweennoise.com/works/2005-2009/transmissions-2005/

Sack, Warren. *The Software Arts*. Cambridge: The MIT Press, 2019.

Schmitt, Philipp. "Camera Restricta." *Philipp Schmitt*. Accessed April 23, 2020. https://philippschmitt.com/work/camera-restricta

Schmitt, Philipp. "Tunnel Vision April 2020." *Unthinking Photography*. Accessed April 30, 2020. https://unthinking.photography/articles/tunnel-vision

Schwab, Katharine. "This Is What Machines See When They Look at Us." *Fast Company*. Accessed April 27, 2020. https://www.fastcompany.com/90139345/this-is-what-machines-see-when-they-look-at-us

Shaken, Edward. "Art in the Information Age: Technology and Conceptual Art." *Leonardo* 35, no. 4 (2002): 433–438.

Striphas, Ted. "Algorithmic Culture." *European Journal of Cultural Studies* 18, no. 4–5 (2015): 395–412.

Sutton, Benjamin. "An Artwork Created by AI Sold for £40,000 at Sotheby's, Failing to Generate the Fervor that Propelled Another AI Work to Sell for 40 Times its Estimate Last Year." *Artsy*. Accessed April 30, 2020. https://www.artsy.net/news/artsy-editorial-artwork-created-ai-sold-40-000-sothebys-failing-generate-fervor-propelled-ai-work-sell-40-times-estimate-year

Taylor, Grant D. *When the Machine Made Art: The Troubled History of Computer Art*. New York: Bloomsbury Academic, 2014.

The Next Rembrandt. Accessed April 24, 2020. https://www.nextrembrandt.com

Verostko, Roman. "The Algorists." *Roman Verostko*. Accessed on April 24, 2020. http://www.verostko.com/algorist.html

Vincent, James. "How Three French Students Used Borrowed Code to Put the First AI Portrait in Christie's." *The Verge*. Accessed April 22, 2020. https://www.theverge.com/2018/10/23/18013190/ai-art-portrait-auction-christies-belamy-obvious-robbie-barrat-gans

Watz, Marius. "The Algorithm Thought Police." *Marius Watz*. Accessed April 18, 2020. http://mariuswatz.com/mwatztumblrcom/the-algorithm-thought-police.html

Index

About the Contributors

Joel S. Beatty is visiting assistant professor of Professional Writing at Fairfield University. His research examines the impacts of visualization technologies on communication, identity and sociotechnical cultures. His work has been published in the edited volume on *Surveillance, Race, and Culture* as well as in the *Community Literacy Journal.*

Ravi Sekhar Chakraborty is doctoral research scholar at the Department of Humanities and Social Sciences of the Indian Institute of Technology Delhi. He works on the philosophical implications of applying formalisms to study literature. More broadly, he is interested in the connection between the philosophy and history of science on one hand and the philosophy and history of politics and literature on the other hand. He has written on the philosophical connections between the thought of L. E. J. Brouwer and Mikhail Bakhtin and is editing an anthology on disciplinarity in the digital age.

Soonkwan Hong is associate professor of marketing at Michigan Technological University. His research focuses on sociocultural and ideological aspects of consumption, which should facilitate our understanding of a variety of consumption practices, consumers' lived experiences, and stylization of their lives. Currently, he studies alternative sustainable lifestyles in multiple ecovillages in the United States and Europe and conducts research on eco-tourism. His research interests also extend to globalization of popular culture. He has published in international journals, such as *Journal of Business Research, Marketing Theory, Qualitative Market Research, Journal of Marketing Management, and Arts and the Market*, and regularly presented at prestigious conferences.

Stefka Hristova is associate professor of Digital Media at Michigan Technological University. She holds a PhD in Visual Studies with emphasis on Critical Theory from the University of California, Irvine. Her research analyzes the digital visual cultures of war and displacement. Hristova's work has been published in journals such as *Visual Anthropology, Radical History Review, tripleC: Communication, Capitalism & Critique, Surveillance and Security, Interstitial, Continuum, Transformations.*

Reka Patricia Gal is a PhD student in Information Studies at the University of Toronto and a fellow at the McLuhan Centre for Culture and Technology working under the guidance of Dr. Sarah Sharma. She holds an MA in Cultural Studies from Humboldt University in Berlin. Her work unites feminist media theory and postcolonial studies with the history of science and environmental studies and explores how technological tools and scientific methods are employed to purportedly solve sociopolitical problems.

Amanda K. Girard is Computer Science Junior Design Capstone Course coordinator at the Georgia Institute of Technology. In her role, she co-teaches and coordinates an interdisciplinary curriculum with the Division of Computing Instruction (DCI) and the Writing and Communications Program (WCP). The curriculum focuses on workplace practice common to Computer Science, such as agile development and user based design. The students work in teams to produce a Minimal Viable Product for an external client, who initially proposed a project idea, and culminates with a public facing expo. Previously, Girard was a Postdoctoral Marion Brittain Fellow in the Writing and Communications Program at Georgia Tech.

James MacDevitt is associate professor of Art History and Visual & Cultural Studies at Cerritos College, as well as the director/curator of the Cerritos College Art Gallery. In addition to founding the multi-institutional *SUR:biennial* and the collaborative *Art+Tech Artist-in-Residence* Program, MacDevitt has curated numerous exhibitions for the Cerritos College Art Gallery, including *OVER/FLOW: Horror Vacui in an Age of Information Abundance*, *Object-Orientation: Bodies and/as Things*, *Architectural Deinforcement: Constructing Disaster and Decay*, *Abstracted Visions: Information Mapping from Mystic Diagrams to Data Visualization*, and *Geo-Ontological: Artists Contemplating Deep Time.*

Ushnish Sengupta is PhD candidate with an expected graduation in 2020 at the Ontario Institute for Studies in Education. Sengupta's research engages with questions of diversity in organizational culture. He has published articles and book chapters that engage with diversity in the field on Organizational Diversity and Political Economy.

Jennifer Daryl Slack is distinguished professor of Communication and Cultural Studies and founding director of the Institute for Policy, Ethics, and Culture at Michigan Technological University. She and Stefka Hristova co-founded the Algorithmic Culture Working Group at Michigan Tech. Professor Slack's research addresses crucial features of contemporary culture from a cultural studies perspective and has both theoretical and practical components. She has published extensively on technology and culture; environment and culture; creativity and culture, and on theoretical issues in cultural studies.

CPSIA information can be obtained
at www.ICGtesting.com
Printed in the USA
BVHW042256050121
597052BV00007BA/22